Systems Failure Analysis

Joseph Berk

ASM International®
Materials Park, Ohio 44073-0002
www.asminternational.org

First printing, December 2009
Second printing, November 2018

Prepared under the direction of the ASM International Technical Book Committee (2008–2009), Lichun L. Chen, Chair.

ASM International staff who worked on this project include Scott Henry, Senior Manager of Product and Service Development; Steven R. Lampman, Technical Editor; Ann Britton, Editorial Assistant; Bonnie Sanders, Manager of Production; Madrid Tramble, Senior Production Coordinator; and Diane Whitelaw, Production Coordinator.

Library of Congress Control Number:2009935432
ISBN-13: 978-1-61503-012-5 (print)
ISBN-10: 1-61503-012-3 (print)
ISBN: 978-1-61503-137-5 (pdf)

SAN: 204-7586

ASM International®
Materials Park, OH 44073-0002
www.asminternational.org

Printed in the United States of America

This book is dedicated to

Chief Gunner's Mate Lawrence F. Costantine

Contents

CHAPTER 1 Systems Failure Analysis Introduction 1
CHAPTER 2 Downsizing the Hidden Factory 11
CHAPTER 3 Systems and Systems Failure Analysis Concepts .. 19
CHAPTER 4 Identifying Potential Failure Causes 25
CHAPTER 5 Fault-Tree Analysis 35
CHAPTER 6 Fault-Tree Analysis Special Topics 47
CHAPTER 7 Fault-Tree Analysis Quantification 57
CHAPTER 8 Failure Mode Assessment and Assignment....... 67
CHAPTER 9 Pedigree Analysis 75
CHAPTER 10 Change Analysis 81
CHAPTER 11 Analytical Equipment 93
CHAPTER 12 Mechanical and Electronic Failures 109
CHAPTER 13 Leaks................................... 131
CHAPTER 14 Contaminants 141
CHAPTER 15 Design Analysis............................ 147
CHAPTER 16 Statistics and Probability 157
CHAPTER 17 Design of Experiments 171
CHAPTER 18 Corrective Action 183
CHAPTER 19 Post-Failure-Analysis Activities 189

APPENDIX A: Component Failure Causes and Suggested Analysis Methods.............................. 193

Index.. 195

CHAPTER 1

Systems Failure Analysis Introduction

THIS BOOK FOCUSES ON solving systems failures. Other books about failure analysis address component or material failures. Component failures are relatively easy to address (e.g., resistors can fail open or sometimes short circuit; metal parts can fail in fatigue, in tension, or in other discernable failure modes; plastic components can experience brittle fractures; etc.). If a capacitor fails, it is fairly simple to cut into it, examine it under magnification, and determine if it was subjected to too much energy (indicated by molten areas) or excess shock or vibration (indicated by mechanical separations).

While portions of this book cover component failure mechanisms, the focus here is on what can cause a system to fail. System failures can be induced by component failures (such as the ones mentioned previously), or they can occur as a result of complex component and subsystem interactions (without any parts failing). Finding the root causes of systems failures is far more difficult.

A few examples illustrate the nature of this challenge:

- When the United States lost the *Challenger* space shuttle in 1986, there was little evidence initially. The team investigating that accident had to evaluate thousands of potential causes and deduce the most likely cause.
- During the 1990 Gulf War, smart munitions received much of the credit for driving Saddam Hussein out of Kuwait. Smart munitions used laser target designators to guide them to their targets. The United States delayed Gulf War action for several weeks, however, because a key laser targeting system could not meet its accuracy requirements. All of the parts in this system conformed to their engineering requirements. No parts failed, but the system did not meet its requirements.

- A large municipal water treatment system designed to remove arsenic from public drinking water lowered arsenic levels to government-mandated requirements, but the system periodically experienced contaminant breakthroughs that allowed high contaminant levels to enter the water distribution system. No parts failed on this system either, yet the system suddenly stopped meeting its requirements.
- Aerospace engineers designed the Apache helicopter blade (a bonded stainless steel, carbon fiber, and titanium structure) to withstand a direct hit from high-explosive enemy projectiles. During combat operations over Iraq, an Apache helicopter blade actually took a direct hit from a Russian-designed ZSU-23/4 high-explosive warhead and returned home. Although the blades can withstand such punishment, their service life in normal operation was significantly less than required by the Army, and the rejection rate during production approached 50%.
- A warehouse storage system relied on wheeled iron structures to increase packing density in pallet storage racks. After thousands of these systems were built, the wheels started failing, requiring an expensive retrofit.

Some system failures are induced by component failures (for example, the warehouse storage system in which the pallet rack wheels failed, or the *Challenger* accident in which an O-ring failed). Some involve interfaces between parts (for example, the Apache helicopter blade failures in which the bonded interface failed). In other system failures, no single component fails, yet the system fails (for example, the laser targeting system or the municipal water treatment system).

Some systems failures are "showstoppers" (the production line comes to halt or a product fails dramatically in service). Others are recurring failures that are not as dramatic as the showstoppers, but they can be very expensive (for example, the high helicopter blade rejection rate during manufacture).

The challenge of systems failure analysis is to define the problem, identify the cause of the problem, select appropriate corrective actions, and then implement the corrective actions. Many organizations get that first step wrong. They do not adequately define the problem. If this is not done correctly, it is highly unlikely the failure analysis team will fix the problem.

In addition, training on systems failure analysis can be helpful at many levels in technical and manufacturing organizations. In addition to becoming knowledgeable in failure analysis procedures, tools, and technologies, failure analysis training instills a way of thinking that helps engineers, manufacturing specialists, purchasing specialists, field service technicians, and quality-assurance personnel become more successful.

The failure analysis process outlined in this book includes the following steps:

- Designate a failure analysis team with representatives from engineering, quality assurance, manufacturing, purchasing, and field service.

- Gather all related failure information.
- Review the aforementioned information and define the problem.
- Identify all potential failure causes using brainstorming, mind-mapping, Ishikawa diagrams, flow charting, the "five whys" technique, or fault-tree analysis.
- List each potential failure cause in the failure mode assessment and assignment.
- Use appropriate documentation reviews, interviews, design analyses, hardware analyses, and designed experiments to converge on the root cause.
- Identify potential interim and long-term corrective actions, and select the most appropriate corrective actions.
- Ensure corrective actions are implemented in all relevant areas (suppliers, inventory, work in progress, repair centers, and fielded systems).
- Follow-up after corrective actions have been implemented to assess corrective action efficacy.
- Evaluate other potential failure causes as corrective action candidates, and incorporate preventive actions where it makes sense to do so.
- Incorporate failure analysis findings into a failure analysis library, design and process guidelines, and troubleshooting and repair documents.

The Mast-Mounted Sight Challenge

One may think that defining the problem is an obvious and simple first step, but that is not always the case. Consider the mast-mounted sight (MMS) system developed and manufactured by McDonnell Douglas. The MMS looks like a basketball (Fig. 1.1) that sits on top of the helicopter blades. That basketball contains a television, a thermal imaging sensor, and a laser

Fig. 1.1 The mast-mounted sight (MMS) on a U.S. Army OH-58 Kiowa helicopter. The MMS contains a laser, a television, and an infrared sensor. Photo courtesy of U.S. Army Aviation Center, Fort Rucker, AL

target designator and rangefinder. Pilots use the MMS for finding targets with either a television or infrared sensor. When the target has been found, the MMS laser determines the range to the target and illuminates the target for the smart munitions.

The MMS laser and television use a window made of glass to see outside the MMS housing. The MMS thermal imaging sensor uses a separate germanium window. (The thermal imaging sensor infrared technology requires a window made of a different material due to its infrared energy requirements.) The MMS contains a boresighting system to align the thermal imaging sensor line of sight with those of the television and laser sensors. This is a critical part of the system and figures prominently in the following discussion.

McDonnell Douglas had been producing and delivering MMS systems to the U.S. Army for years. Prior to actually going to war, however, the Army decided to thoroughly wring out its MMS systems. During precombat testing, the Army found that the laser beam was misaligned enough to induce a miss, and this condition existed on all of its MMS systems. This was a huge problem. It threatened the United States' ability to go to war.

The engineers assigned to solve this system failure jumped to a conclusion and immediately defined the problem as laser misalignment. The failure analysis team attributed the cause to an MMS boresighting failure, and they spent several months attempting to find the failure cause in the boresighting portion of the system. These were smart people (the failure analysis team included engineers, physicists, scientists, and others), but they missed the first step of the problem-solving process: They did not define the problem correctly. To compound the situation, without attempting to identify all potential causes, the failure analysis team jumped to another conclusion when they decided that the cause must be in the boresighting system. The failure analysis team lost valuable time looking for the failure there.

After several months without making any progress, the McDonnell Douglas engineers and scientists decided to re-examine the test data. When they did this, they made two critical discoveries. The problem only appeared at cold temperatures, and both the laser beam and the television sensor were misaligned with the thermal imaging sensor. The laser beam and the television sensor were in alignment with each other. In other words, the problem was not laser beam misalignment with both of the other sensors, as the team previously thought. The team now recognized that the television and the laser were aligned with each other, but both were misaligned with the thermal imaging sensor. This was a different problem than the one the team had been attempting to solve.

Armed with this new (but previously available) information, the failure analysis team looked beyond the boresighting system. In particular, the team identified what part of the MMS system operated differently in cold weather. The team recognized that the laser and television sensor window used a different window-heating system than did the thermal imaging sensor. They next found that these heaters operated at different temperatures. Within hours, the failure analysis team concluded that the window heaters had induced the cold weather misalignment. McDonnell Douglas specified a simple software fix to correct the problem, and the Army implemented it on a worldwide basis a few days before engagement in the liberation of Kuwait.

The lessons inherent to the aforementioned experience are to ensure that the problem is accurately defined, that conclusions are not hastily made, and that all potential causes are considered before a fix is attempted.

The Systems Failure Analysis Process

The challenge in approaching a system failure is to:

- Accurately define the problem
- Identify all potential failure causes
- Objectively evaluate the likelihood of each failure cause
- Take steps to prevent these failure causes from occurring

This is an extremely important concept, so much so that it is highlighted as follows:

> *The systems failure analysis approach requires defining the problem, identifying all potential failure causes, objectively evaluating each potential failure cause, and implementing actions to preclude recurrence.*

This approach works well for several reasons.

Focusing on identifying all potential failure causes (without arbitrarily or subjectively eliminating any during the initial analysis phase) opens a universe of potential failure causes. These probably would not be considered if the failure analysis team jumped to and addressed only the most likely causes. Several techniques for identifying all potential failure causes are covered. For now, it is important to recognize that the objective is to identify all potential causes, not just the perceived obvious ones.

If the failure being analyzed is a recurring or intermittent condition, the failure causes will almost certainly be subtle. Identifying all potential causes forces the investigator to look away from the obvious causes. If the cause of a recurring or intermittent problem was obvious, would not prior failure analysis efforts have already identified and corrected it?

When the failure analysis team focuses on identifying all potential failure causes, the failure analysis team will identify potential causes beyond those that caused the failure under investigation. Even if the failure analysis team determines that these other hypothesized failure causes did not cause the failure being investigated, this approach creates numerous improvement opportunities. The failure analysis team can address the other hypothesized causes and prevent them from recurring as well.

The Four-Step Problem-Solving Process

In subsequent chapters, this book introduces and develops several sophisticated approaches for identifying and evaluating potential failure modes, developing potential corrective actions, and then selecting the best corrective actions. All of these can be condensed, however, to the simple four-step problem-solving process shown in Fig. 1.2.

Each of these steps is examined as follows.

What Is the Problem? Defining the problem sounds easy. It frequently is not. Based on experience in hundreds of organizations spanning several industries, this is a step that many people miss (consider the McDonnell Douglas MMS example described previously). It is very easy to focus on symptoms or to jump to conclusions regarding potential causes and thus miss the problem. Therefore, it is highly recommended to spend enough time on this step. All members of the failure analysis team should agree that the problem has been accurately defined before moving on to the next step.

What Is the Cause of the Problem? After defining the problem, the failure analysis team can use several technologies to identify potential failure causes. It is important to recognize that this is not a simple process. It is also important to realize that this question is not always treated as objectively as it should be. Consider these scenarios:

- One or more of the participants in a failure analysis meeting feels confident that they know what caused the failure before all of the facts are available.
- Potential failure causes are dismissed without careful consideration.
- The people in such discussions jump ahead to define corrective actions before the failure causes have been confirmed.

During this step of the four-step problem-solving process, the failure analysis team should focus on accomplishing two objectives. The first is to identify all potential failure causes. The second is to objectively evaluate the likelihood of each. This book develops a structure for doing both.

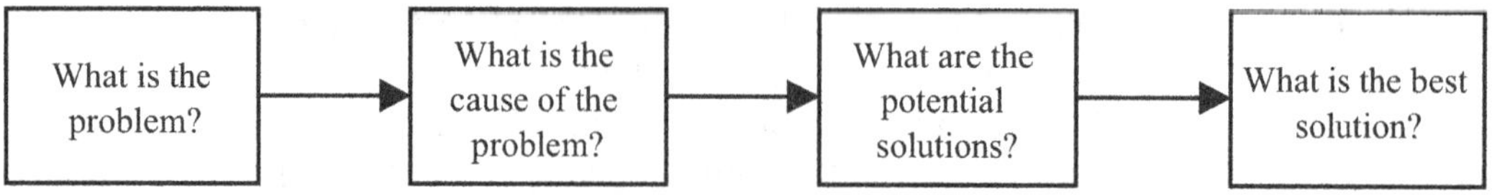

Fig. 1.2 The four-step problem-solving process guides the systems failure analysis technologies and approach addressed in the remainder of this book.

To help identify all potential failure causes, the following methods are covered:

- Brainstorming
- Ishikawa diagrams
- The "five whys" approach
- Mind mapping
- Fault-tree analysis

All of these approaches for identifying potential failure causes are good ones, but fault-tree analysis is preferred in many cases for its systematic coverage. After the failure analysis team has identified all potential failure causes, the team focus should then shift to objectively evaluating each. This book describes several technologies covering this important part of the systems failure analysis process.

What Are the Potential Solutions? Identifying the potential solutions can only occur after the causes have been identified, and there are categories of corrective actions ranging from highly desirable to those that are not so desirable. Highly desirable corrective actions are those that remove all potential for human error. These corrective actions can include such things as designs that prevent incorrect assembly or processes with capabilities that guarantee outputs within acceptable ranges. Less desirable corrective actions rely on people doing things correctly. Examples of these include additional inspections to eliminate defects (an approach highly likely to allow defects to escape, as is explained in later chapters), assembly sequences that can induce defects if not followed exactly, and other actions similarly dependent on people doing things perfectly.

What Is the Best Solution? The aforementioned guidance notwithstanding, the best solution may not always be the one selected for immediate implementation. Sometimes, product or process redesigns are not feasible for cost or schedule reasons. Sometimes, additional inspections or screenings are the only avenue available in the short term. In many cases, the failure analysis team may opt to implement interim, less desirable corrective actions immediately, to be followed up by longer-term, more desirable corrective actions that absolutely preclude failure recurrence.

As mentioned previously, the failure analysis team should not restrict its thinking only to the actual failure cause. The failure analysis team should also consider corrective actions to prevent the other hypothesized causes from inducing future failures. The failure analysis team should also go beyond the system that failed. In many cases, other identical systems will have already been fielded or placed in storage. Other systems may be in production. The failure analysis team should evaluate all product areas to determine if these areas should incorporate the corrective actions applied to the failed system. Finally, the failure analysis team should consider other similar system designs susceptible to the same problems. Where appropriate, these also receive the same corrective actions.

The Failure Analysis Team

Many organizations assign failure analysis responsibilities to a single department (typically engineering or quality assurance). This is a mistake. Experience has shown that the most effective failure analysis teams include engineers, quality-assurance specialists, manufacturing technicians, purchasing personnel, field-service personnel, and others. The inclusion of an organization's major disciplines ensures that no single department subjectively and unilaterally concludes the fault lies outside their area of responsibility. A system design, manufacturing process, tooling, or inspection approach can induce failures. Components or subassemblies provided by suppliers can induce failures. The environment in which the system is operated can induce failures. There are many other factors that can induce failures. For example, manufacturing organizations typically purchase more than half of their product content; including a purchasing representative on the team ensures quick and accurate communication with suppliers. The failure analysis team will need to assess the product pedigree, which may require additional testing or inspection; including a quality-assurance representative will expedite obtaining this information. In some cases, failures occur even when all parts are conforming to the engineering drawings and the system has been properly assembled; in such cases, it makes sense to have an engineer on the team to assess design adequacy.

There is a synergy that emerges when a failure analysis team composed of different specialists defines the problem, identifies and evaluates potential failure causes, develops potential corrective actions, and selects the best corrective actions. Incorporating representatives from each critical area fosters problem-solving synergy and ensures that the organization's strengths and capabilities are appropriately focused.

Summary

The four-step problem-solving process is a basic framework for systems failure analysis. The failure analysis team should start by gathering all available information and converge on a clear, agreed-upon problem definition. The next step is to identify all potential failure causes. When the potential failure causes have been identified, the failure analysis team should objectively evaluate each. This will guide the failure analysis team to the cause of the system failure under investigation. The failure analysis team should evaluate and implement corrective actions for the confirmed cause of the failure and for other potential failure causes, to prevent other future failures. The failure analysis team should include engineering, manufacturing, quality assurance, purchasing, field service, and representatives of other disciplines to ensure problem-solving synergy and objectivity.

Example for Group Discussion. Military range-finding and target-designation lasers (such as the MMS) typically use a 1.06 μm wavelength laser beam, which is hazardous. The 1.06 μm wavelength laser beam can permanently blind a human. For this reason, some military lasers convert the laser energy to a nonhazardous 1.54 μm wavelength for training exercises. The LANTIRN laser system accomplishes the wavelength shift with the use of a device called a Raman cell (named for the scientist who discovered the effect). The Raman cell is a sealed titanium tube containing 1000 psi methane gas and windows at either end. The Raman cell converts 1.06 μm wavelength laser energy to a 1.54 μm wavelength.

During production of the LANTIRN laser assembly, technicians direct the 1.06 μm wavelength beam into the Raman cell. The assembly technicians then adjust the amount of energy entering the cell to meet required Raman cell output energy levels, because some energy is lost when the beam travels through the cell.

During early LANTIRN production, approximately 10% of the Raman cells failed when the technicians adjusted the laser energy. When the technicians adjusted the energy level, some of the Raman cell windows developed dark-brown burn spots. These burn spots can result from:

- The windows being contaminated (i.e., they are dirty)
- Too much laser energy passing through the windows
- A combination of too much laser energy and contamination

The laser system manufacturer accepted the 10% rejection rate, because they considered this to be part of the "black art" of laser manufacturing. After several months, however, the Raman cell rejection rate due to burn spots crept upward until it hit 50%. The manufacturer could not accept a rejection rate this high.

Based on the aforementioned:

- How should the problem be defined?
- What are the causes of the problem?
- What are the potential solutions?
- What is the best solution?

SELECTED REFERENCES

- *Aircrew Training Manual: OH-58D Kiowa Warrior,* Department of the Army, 2000
- S. Berk and J. Berk, *Quality Management for the Technology Sector,* Newnes, 2000
- P.C. Reid, *Well Made in America: Lessons from Harley-Davidson on Being the Best,* McGraw-Hill Publishing Company, 1990
- L. Stewart and J. Berk, Ordnance System Failure Analysis, *Proceedings of the Annual Reliability and Maintainability Symposium,* 1987

CHAPTER 2

Downsizing the Hidden Factory

NEARLY EVERY ORGANIZATION has experienced "showstopper" system failures, similar to the one described for the mast-mounted sight. These are the failures that bring an organization (or its customers) to a halt. Other failures in this category may include:

- An airplane departs from controlled flight and crashes.
- A computer system shuts down and prevents communication, work, and so on.
- A water treatment system allows contaminant levels to go above government-specified levels.

These and other such failures are dramatic in that they have immediate and unacceptable consequences, and they focus the organization's attention on finding the causes and fixing the problem. Their impacts are serious and immediate. They cannot be ignored.

Organizations also can experience less dramatic, recurring failures that are not showstoppers. These can also be significant. Although these significant, show-stopping failures get attention, they are usually not the most expensive problems. That distinction goes to the hidden factory. The costs associated with scrap and rework can be huge, especially when considered over time. Consider the LANTIRN laser Raman cell example at the end of Chapter 1. This was a manufacturing process yield problem. The manufacturer in that case tolerated it at 10% but found it unacceptable at 50%.

Many organizations are willing to accept yields less than 100% (sometimes significantly so). The organization may consider the defects to be a normal part of the manufacturing process, and they are accepted as a cost of doing business. Sometimes, warranty claims are regarded similarly. If the costs are below some historical or other level that management has decided is acceptable, the costs are accepted as inherent to the business. This is not a good idea.

The tools of systems failure analysis discussed in this book can be applied to recurring problems of the hidden factory, just as well as show-stopping failures that rivet the organization's attention. The tools of systems failure analysis can provide returns over time, when the tools are used to find and correct root causes of unacceptable variations in any type of system.

The Hidden Factory

The hidden factory refers to the activities associated with scrap and rework. Management can sometimes stop seeing scrap and rework as the cash drain that it is. In a very real sense, management is paying for the work twice—once when the work is accomplished the first time, and then again when the work is either scrapped and built again or when it is done again as rework. This is called the hidden factory. It is not actually hidden in the sense that it is done in a concealed or secretive manner. It is in plain view but is not recognized as another manufacturing center.

The costs associated with scrap and rework can be huge. In a motor home manufacturing facility that employed approximately 400 people, a third of the workforce was assigned on a regular, full-time basis to correct cosmetic and functional defects at the end of the assembly line. This was after all work had been accomplished and accepted at earlier steps in the manufacturing process. This manufacturer could have reduced its labor content by one-third of its labor force (approximately 130 full-time employees) if the motor homes did not require rework at the end of the line.

Although the motor home example is blatant, its rework costs are not unusual. Various estimates of the cost of poor quality consistently place these costs at 15 to 50% of the total cost of doing business (Ref 1, 2). One can argue about the exact amount, but the important thing to note is that such costs are significant.

The name notwithstanding, and even though it is usually diffused throughout the organization, finding the hidden factory is easy. It simply requires looking for rejections, scrap, and rework anywhere in the process. The following are suggested places to look:

- Scrap bin (Examine its contents.)
- Warranty returns area (Examine items that are being returned for warranty rework or replacement.)
- Rejected item storage areas (Look for items that have been rejected.)
- Shipping dock (Look for items that are being returned to suppliers rather than being shipped to customers.)
- Each workstation (Look for items that have been returned or retained for rework.)

The aforementioned provides a rudimentary understanding of what sorts of things the hidden factory is doing, and sometimes, that is enough to get started. At some point, more quantitative hidden factory measures are necessary. Those are covered shortly.

Quantifying the Hidden Factory

Vilfredo Pareto, an Italian economist who lived from 1848 to 1923, observed that 80% of Italy's wealth was concentrated in 20% of Italy's population (in other words, a few people had most of the money). This observation became known as the 80/20 rule, or the Pareto principle. The Pareto principle can be applied to many things:

- Good salesmen intuitively make most of their sales to only a few of their customers.
- Most stocks are traded by a relatively small number of stock buyers and sellers.
- Most beer is consumed by a small percentage of the people who drink beer.

The Pareto principle also works well in the manufacturing world. Most organizations find that the bulk of their rejections are attributable to a small percentage of the rejection causes. Finding and correcting these dominant rejection causes can eliminate a disproportionately large number of defects. The objective is to identify which defects occur most frequently, or cost the most, or have the most profound impact on the business or its customers. For the moment, that last category can be ignored, because those tend to be the showstoppers previously discussed. The showstoppers automatically get attention. This chapter focuses on how to get one's arms around the rejections that occur most frequently or have the highest cost.

The best way to do this is to collect rejection data on all nonconformances. Some organizations have sophisticated quality-measurement systems to do this; others do not. While sophisticated quality-measurement systems make it easy to prepare a Pareto analysis of the most frequently occurring or most expensive nonconformances, anyone with access to Microsoft Excel can do this easily.

Suppose a shelving manufacturer wants to prepare a Pareto analysis to identify dominant recurring nonconformances from both frequency and cost perspectives. The manufacturer can collect and enter the rejection data for the preceding month into an Excel spreadsheet and then sort it, as shown in Tables 2.1 and 2.2.

Pareto data are typically organized into a bar chart, showing the most frequently occurring defect on the left side of the chart, and then showing bars for each defect to the right in descending order of occurrence. In the

Table 2.1 Quantity of shelf manufacturing defects in the preceding 30 days

Defect	Frequency during preceding 30 days
Shelf too short	156
Shelf too long	48
Hole not drilled	687
Missing paint	299
Weld porosity	435
Paint runs	382

Table 2.2 Quantity of shelf defects during the preceding 30 days, sorted by most frequently occurring to least frequently occurring

Defect	Frequency during preceding 30 days
Hole not drilled	687
Weld porosity	435
Paint runs	382
Missing paint	299
Shelf too short	156
Shelf too long	48

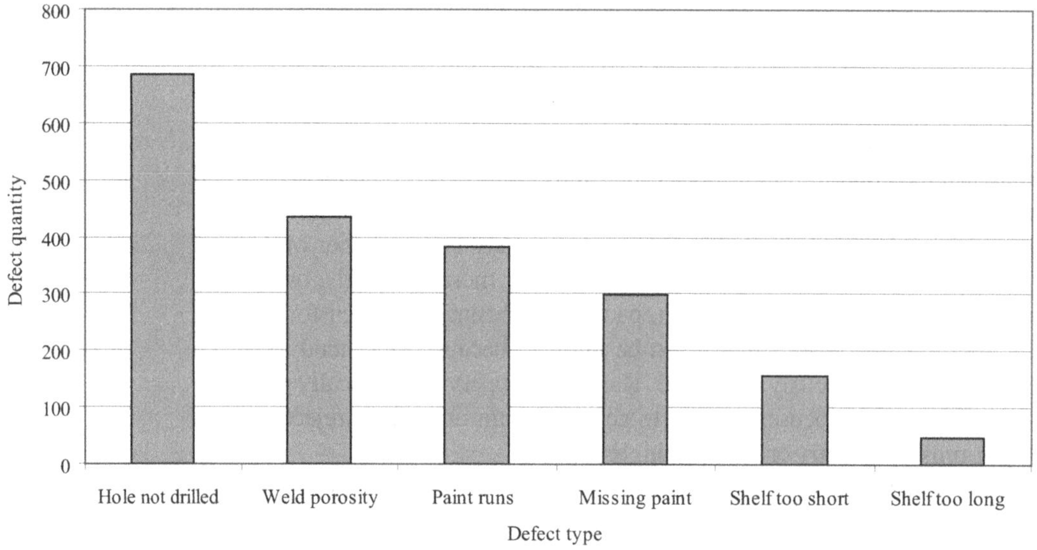

Fig. 2.1 Shelf manufacturing quatity-based Pareto analysis showing defect quantities arranged by frequency during the preceding 30 days

case described previously, the shelving manufacturer can use the Excel charting feature to prepare the Fig. 2.1 Pareto chart.

The quantity-based Pareto chart in Fig. 2.1 shows the most frequently occurring nonconformance, the next most frequently occurring nonconformance, and so on. This Pareto chart shows that "hole not drilled," "weld porosity," and "paint runs" are the most frequently occurring nonconformances. This suggests that if the manufacturer wishes to reduce the defect frequency, attention should be focused in these areas.

Table 2.3 Manufacturing defect cost data for each component, the value added to the component by the time it is rejected, and the total cost of each rejection category

Defect	Frequency during preceding 30 days	Component cost	Value added	Total cost
Hole not drilled	687	$3.32	$6.00	$6,402.84
Weld porosity	435	$11.04	$13.96	$10,875.00
Paint runs	382	$13.87	$5.86	$7,536.86
Missing paint	299	$2.04	$2.04	$1,219.92
Shelf too short	156	$19.05	$8.87	$4,355.52
Shelf too long	48	$5.04	$8.87	$667.68

Table 2.4 Manufacturing defect cost data, sorted by total cost, for each defect, sorted from most costly to least costly

Defect	Frequency during preceding 30 days	Component cost	Value added	Total cost
Weld porosity	435	$11.04	$13.96	$10,875.00
Paint runs	382	$13.87	$5.86	$7,536.86
Hole not drilled	687	$3.32	$6.00	$6,402.84
Shelf too short	156	$19.05	$8.87	$4,355.52
Missing paint	299	$2.04	$2.04	$1,219.92
Shelf too long	48	$5.04	$8.87	$667.68

As mentioned earlier, the manufacturer can also prepare a Pareto chart to show the most costly defects. The manufacturer can add data to the defect summary showing the cost of each component and the value added to each component at the time it is rejected, as shown in Table 2.3.

Table 2.3 shows that when total costs are considered, the defects are no longer rank-ordered. Note that in Table 2.3, the manufacturer must consider the value added to each component at the time the defect is discovered, because those are costs that will be lost when the item is rejected. The manufacturer can sort the table again to rank-order the defects by total cost, as shown in Table 2.4.

Note that when considered from the cost perspective, the most significant items move around a bit. From the cost perspective, "weld porosity," "paint runs," and "hole not drilled" become the most significant defects. Figure 2.2 shows the comparable Pareto chart.

Summary

The hidden factory represents scrap and rework. The rework and scrap activities are visible, but manufacturers frequently accept the associated costs as inherent to the process. For that reason, these activities are referred to as the hidden factory. The costs associated with hidden factory activities are significant over time. For this reason, applying systems failure analysis to reduce or eliminate the hidden factory makes sense. Pareto analyses can be used from both frequency-of-occurrence and cost perspectives to target specific areas for improvement.

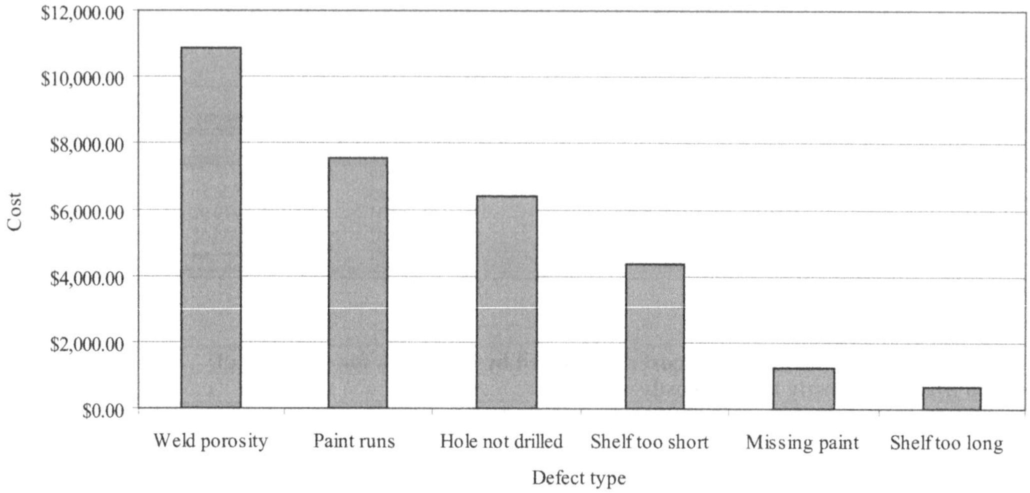

Fig. 2.2 Shelf manufacturing cost-based Pareto analysis showing defect costs for the most expensive defects during the preceding 30 days

Example for Group Discussion. The motor home manufacturer example cited at the beginning of this chapter produced 20 coaches every day. The assembly line had approximately 15 workstations. Inspectors identified defects in each workstation before the coaches moved to the next workstation, and each workstation reworked the coach to eliminate or correct the defect prior to the coach moving to the next workstation. At the end of the assembly line, the coaches underwent a final inspection, where the inspectors identified remaining defects by applying pieces of red tape to the defect areas. As mentioned earlier, a third of the production work force (approximately 130 people) corrected the defects noted in the final inspection area. After this rework activity was complete, the coaches were shipped to recreational vehicle dealers. The motor home manufacturer received numerous quality complaints from dealers and from people who purchased the motor homes.

The motor home manufacturer's president was quite concerned about quality. He and the marketing vice president randomly selected three ready-to-ship coaches every day. They personally inspected these three coaches. They felt that by doing this they communicated quality commitment to the rest of the factory. When asked, the president admitted that he had been doing this for several months, and the approximate number of defects he and the marketing vice president discovered each day remained the same.

What comments do you have about the president's approach to demonstrate a commitment to quality (i.e., his daily inspection of three motor homes)?

What do you think about the company's overall inspection approach and its effectiveness?

If you were the president of this company, what would you do?

REFERENCES

1. L. Sorqvist, Identifying the Cost of Poor Quality, *Measuring Business Excellence*, Vol 2, 1998
2. H.J. Harrington, *Poor Quality Costs,* Marcel Dekker, Inc., 1987

SELECTED REFERENCE

- D.C. Wood, *The Executive Guide to Understanding and Implementing Quality Cost Programs,* ASQ Quality Press, 2007

CHAPTER 3

Systems and Systems Failure Analysis Concepts

A SYSTEM IS A COMBINATION of parts, subassemblies, procedures, and people that form an entity capable of accomplishing specific functions when commanded to do so. It may be to transport people or freight, it may be to launch a projectile, or it may be to manufacture a product.

Consider these examples:

- The automobile is a transportation system. This system is a collection of both parts (the various components that make up the automobile, such as pistons, rings, battery, tires, etc.) and things (the driver, the software that controls the engine management system, and even the roads, traffic lights, gas stations that provide fuel, and other items). The purpose of this system is to transport people from one point to another, but that is not its only purpose. An automobile must stop within a reasonable distance, accelerate up to highway speeds, provide radio reception, travel in the direction the driver commands it to, secure things in the trunk, and perform other functions.
- The personal computer is a system consisting of parts, subassemblies, and software. The personal computer has many purposes. Depending on how the computer is configured, it can allow its operator to write and print reports, perform mathematical operations, generate engineering drawings, access the internet, and do many other things.
- A manned aircraft is another transportation system. A manned aircraft is comprised of many thousands of parts, hundreds of subassemblies, people to operate and support it, and software. The purpose of this system is to provide human air travel, but this is only one of its purposes. A manned aircraft also provides an environment in which passengers can survive the journey. Commercial passenger aircraft provide restroom,

entertainment, and food service features. Combat aircraft provide weapon-carrying, aiming, and launching capabilities. Some are designed to take off and land vertically.

- A rifle is a system designed to launch a projectile at high velocity and hit an intended target. Rifles contain mechanical parts that chamber the cartridge, lock the cartridge in place, fire the cartridge, launch the bullet toward the target, and eject the spent cartridge casing. A rifle also includes parts that allow the operator to aim it. The primary purpose of this system is to allow the user to strike a target with a projectile, but the rifle has many other purposes. These other purposes include providing a mount for the rifle sights, carrying additional ammunition, spinning the projectile as it travels through the barrel, allowing the user to make the rifle safe so it cannot be fired if the trigger is accidentally pulled, and many others.
- Even the cartridge mentioned previously is a system consisting of the projectile, the propellant, the primer, and the cartridge case. The purpose of this system is to carry and ignite the propellant to launch the projectile. This simple four-part system also has several functions, such as maintaining its structural integrity, keeping the propellant and primer dry, holding the bullet securely until it is fired, releasing the bullet when the propellant is ignited, forming a seal within the rifle chamber such that no propellant gases are released, and interfacing with the rifle chamber.
- A manufacturing process is also a system. Manufacturing processes bring together tools, techniques, processes, and human beings to form parts and collections of parts that meet multiple dimensional, reliability, quality, appearance, and performance requirements.

Basically, anything consisting of more than one item that is designed to accomplish one or more functions is a system.

Systems Failures

A system failure occurs when a system does not do what it is supposed to do when it is supposed to do it, or it does something it is not supposed to do.

As can be seen in the preceding paragraphs, most systems have many requirements, and they are supposed to do more than just one thing. A system failure occurs if the system fails to meet any of its requirements. For example, a commercial airliner may transport its passengers from one airport to the next, but suppose the aircraft environmental control system is running too hot and the cabin is uncomfortably warm. That is a system failure. Suppose that a rifle fires a cartridge, but the sights are misaligned and the bullet misses its target. That is a system failure. Suppose a manufacturing system produces a part that misses one or more of its dimensional requirements. That, too, is a system failure. In all of these cases, the system did not meet all of its requirements. The system did not do what it was supposed to do. The system failed.

Intermittent System Failures. Many people have experienced intermittent system failures with a car or some other system. The system does not work the way it is supposed to, but when it is taken to a mechanic for repair, it works.

Intermittent failures can be incredibly frustrating for a number of reasons. Sometimes, the failure does not occur when attempts are made to duplicate it, which makes it difficult to isolate the cause. Sometimes, a fix is guessed at and implemented, and the failure goes away. This can lead to a "feel-good" situation, but this feeling is frequently temporary. If something is done to fix an intermittent problem without knowing the failure cause, and the problem goes away the next time an attempt is made to use the system, has anything been proven? The problem is intermittent. It may not occur again even if nothing is done.

The techniques covered in this book for identifying all potential failure causes work well for correcting intermittent system failures. The thoroughness and objectivity of the systems failure analysis approach does not rely on an intermittent failure manifesting itself when the failure analysis team is evaluating potential failure causes.

Inadvertent System Function. The previous system failure definition mentioned the case in which a system does something when it is not supposed to. This is also a system failure. Such occurrences can range from events that are mildly amusing to those that are deadly. A few decades ago, an automobile manufacturer built cars that occasionally activated the windshield wipers when the driver turned the radio on (this was caused by a sneak circuit condition, which is covered in a subsequent chapter). The aforementioned windshield wiper/radio sneak circuit situation seems almost comical, but others are decidedly not. A bomb that detonates inadvertently is a system failure with almost certain disastrous consequences.

What if a nuclear intercontinental ballistic missile launched inadvertently? Such an event would be cataclysmic and incomprehensible. It was precisely this hypothesized system failure that led to the creation of the fault-tree analysis technique. The U.S. Air Force realized it had no analytical tool that could adequately define (and thereby allow people to prevent) conditions that would result in such an event. To respond to this gap, the Air Force, Boeing, and Bell Laboratories created fault-tree analysis. Fault-tree analysis is discussed in much greater detail in the next several chapters. It is the basic tool primarily used for answering the second question in the four-step problem-solving process (What is the cause of the problem?).

The technologies discussed in this book are equally applicable to identifying the root causes of inadvertent function system failures.

Understanding How Systems Operate

To understand how a system can fail, the failure analysis team must first understand how the system operates. There are a number of ways to do this, but first, one needs to recognize that knowing how the system is supposed to operate does not mean one will know how it can fail. In fact, systems designers and development engineers (while helpful in defining how the system is supposed to operate) are sometimes not very helpful in defining how it can fail. Designers and development engineers are trained to think in terms of how the system is supposed to operate. It is hard for them to think in terms of how it can fail.

Think about what happens when writing and proofreading. When writing and then proofreading one's own work, are all of the spelling, typographical, and other errors found? The answer is usually no. Proofreading is thus essential in the writing process, because any writer frequently misses mistakes or poor wording. The same phenomenon occurs when system designers and development engineers evaluate failures on systems they designed. They see what the system is supposed to do, not how it can fail.

When the failure analysis team analyzes how a system fails after it has learned how it is supposed to work, the team will know more about the system subtleties than will the designers and development engineers. Good engineers recognize this, and they frequently consult with the people who have analyzed earlier systems failures. Good engineers recognize the differences in perspective between someone who designs the system and someone who analyzes how it can fail, and they want that valuable insight during the design process to eliminate potential failure causes.

There are several approaches for learning how systems work:

- Interviewing the system designers and development engineers is a great place to start. Most systems are complex, and it may be that no single engineer knows how the entire system is supposed to work. The failure analysis team may have to interview mechanical, electrical, and software engineers to learn how the system is supposed to operate.
- More complex systems have specifications defining what the system is supposed to do (especially military equipment or other items procured by large institutional buyers). These specifications frequently provide a systems operation overview.
- Sales literature and proposals frequently include system descriptions.
- Field support, startup, and maintenance personnel have insights into how a system operates. Based on the nature of their work, these people also have insights and first-hand experience in how systems fail.
- The people who actually manufacture, assemble, and test systems are perhaps the best source for understanding how a system operates. These people often have an intuitive feel for the system and how it can fail.

- Customers (especially on complex, custom-engineered products) often understand how systems are supposed to operate, especially because they probably created the system specification. Customers live with the product, and they often have extensive experience with the system.
- Suppliers have excellent insights into how their portions of the system operate. The content of most systems comes from suppliers.

Summary

A system is an assemblage of people and things designed to accomplish specific functions when commanded to do so. Most systems perform multiple functions. System failures occur when the system does not meet all requirements or if the system functions inadvertently.

To analyze how a system can fail, the failure analysis team must first understand how the system is supposed to operate. Systems designers and development engineers should be part of the failure analysis team, but the failure analysis team should recognize that engineers think in terms of how the system operates (not how it can fail). After understanding how the system is supposed to operate and then analyzing how it can fail, the failure analysis team will understand the system nuances better than the people who originally designed it.

Example for Group Discussion. The F-16 Air Combat Fighter, developed in the mid-1970s by General Dynamics, was the world's first "fly-by-wire" aircraft. Instead of mechanical cables moving flight control surfaces, the F-16 used electrical signals routed to servoactuators. The F-16 requirements included freedom from any single-point failure (i.e., a single failure or event) inducing a departure from controlled flight. Because of this, it was critical that no single-point failure result in loss of the electrical commands to the flight control servoactuator. If the signal was lost, the servoactuator would drive the flight control surface to its extreme position, and the aircraft would crash.

To address this concern, the F-16 development engineers designed quadruply-redundant signals to each servoactuator. Any three of the four signals could be lost, and the remaining signal would still control the servoactuator. The engineers' reasoning was that the probability of losing all four signals was extremely remote, and that no single-point failure could induce this condition.

General Dynamics organized a separate group of engineers to analyze the F-16 design. This separate team found that although the F-16 included quadruply-redundant flight control signals, the electrical harnesses carrying these signals used common connector plugs. If any of several hundred connectors disconnected or if the harnesses carrying the signals were cut, all signal paths would be lost. This was a surprise to General Dynamics and to the F-16 design engineers. Correcting the problem and eliminating these single-point failures required significant redesign.

Why did the F-16 development engineers miss the single-point failures described previously?

Why was a separate engineering team able to find the design deficiency?

Only a few F-16s had been built when the aforementioned problem was discovered, and no failures had occurred. This shows that systems failure analysis technologies can be applied before failures occur. What are the advantages of such an approach? Are there any disadvantages?

SELECTED REFERENCES

- *Ballistic Systems Division Total Quality Management Handbook,* Air Force Systems Command, Headquarters Ballistic Systems Division, 1989
- C.S. Droste, "Lessons Learned in the Development of the F-16 Flight Control System," General Dynamics Fort Worth Division, 1983
- C. Kopp, Artificial Stability and Fly-By-Wire Control, *Australian Aviation and Defence Review,* 2005

CHAPTER **4**

Identifying Potential Failure Causes

IN THE SECOND STEP of the four-step problem-solving process, the failure analysis team should identify all potential failure causes. There are several approaches for doing so:

- Brainstorming
- Mind mapping
- Ishikawa diagrams
- The "five whys" technique
- Flow charting
- Fault-tree analysis

Each of the aforementioned techniques is useful for identifying potential failure causes, but fault-tree analysis is, by far, the most powerful tool for this purpose. The next several chapters address fault-tree analysis, which is a far more rigorous analysis approach than any of the previously mentioned techniques and is preferred in many cases. For simpler problems, however, the techniques outlined in this chapter can be useful.

Brainstorming

Brainstorming is a group discussion designed to elicit as many ideas as possible. An advertising executive named Alex Osborn formalized the technique in the 1940s to help develop ideas for selling new products. In its simplest form, brainstorming involves a group engaging in a free-flow of suggestions to solve a problem. The approach typically includes the following steps:

- *Notification:* The person scheduling the brainstorming sessions notifies the people invited to the session, with a clear problem statement.

This allows those invited to the meeting to start thinking about potential solutions prior to attending the meeting.

- *Introductory comments:* The brainstorming session leader states the problem to be solved and the meeting rules at the beginning of the session.
- *Idea generation:* The brainstorming participants make potential problem-cause suggestions.
- *Idea documentation:* The brainstorming session leader captures all suggestions on paper. It is best to do this on large pads on an easel. The session leader can tape completed sheets to the wall so that they remain visible during the session. Keeping prior ideas visible helps to stimulate new suggestions, and it provides a record for the meeting notes.
- *Meeting notes publication:* After the brainstorming meeting, the session leader makes and circulates copies of the meeting notes.

In many cases, the ideal size for a brainstorming session is five or six people. If the group is much smaller than this, the idea-generation process seems to be tougher to start and sustain. If the group is larger, it becomes difficult to keep up with the ideas and to get everyone to participate. It is recommended that people with different backgrounds be included, because this ensures the emergence of ideas from different perspectives. Including an engineer, a manufacturing person, a quality-assurance specialist, a buyer, and so forth on the failure analysis team (as recommended earlier) ensures a mix of backgrounds.

It is not recommended to include people at different management levels, because lower-level people may be intimated by upper-management people, and this can stifle the free generation of ideas. An exception may be the inclusion of a manufacturing technician (usually an hourly, nonmanagerial type). Manufacturing technicians frequently have an intuitive feel for how the product should work, how it may fail, and other nuances that other failure analysis team members may lack.

The meeting should be kept to one hour or less. People tend to become bored in meetings lasting longer than one hour. The meeting should be held in a quiet room with no interruptions. Meetings over lunch generally do not work well, because people are thinking about eating. Everyone should be asked to turn off their cell phones, because the distractions from phone calls, text messages, and e-mail can interrupt the idea-generation process.

In brainstorming sessions, the flow of ideas is usually slow to start. Then, the ideas start to flow freely (so much so that keeping up with documenting the ideas can be difficult). After a while, the process decelerates, and ideas flow much less freely. The idea-generation process seems to follow the same pattern as making microwave popcorn. At first, not much happens, and the kernels seem to pop very slowly. When they start popping, the popcorn occurs rapidly. After a period of time, the process slows, and kernels pop infrequently.

When the flow of ideas slows, the failure analysis team leader can use any of several techniques to reenergize the process. These include:

- *Seeking similar problems and solutions:* The failure analysis team leader can ask the brainstorming participants to consider other similar problems and their solutions.
- *Getting wild:* The failure analysis team leader can ask the team participants to think "outside the box" and develop solutions that appear to be completely outrageous. Sometimes, such thinking evokes solutions that would not have otherwise been considered.
- *Considering other perspectives:* The failure analysis team leader can ask the brainstorming participants to consider looking for solutions from other perspectives.
- *Using a reporter's approach:* The failure analysis team leader can ask the team to consider the problem from a journalist's perspective and ask the questions a reporter would ask (who, what, when, where, why, and how).
- *Mental imaging:* The failure analysis team leader can ask the team to imagine themselves physically in the system and ask (from this new perspective) what may be causing the problem. For example, if a light bulb fails to illuminate, the brainstorming session members can mentally picture themselves in the light bulb interior and ask what could cause the illumination failure.
- *Considering more (or less) is better:* The failure analysis team leader can ask the team to consider previous suggestions from the perspective of adding more (or less) of the suggestion.
- *Seeking synonyms:* The failure analysis team leader can ask the team to identify synonyms for each previously suggested potential cause.
- *Expanding previous ideas:* The failure analysis team leader can ask the team to consider previous suggestions that emerged during the brainstorming session and expand on them.
- *Rereading ideas:* The failure analysis team leader can reread all of the previous suggestions aloud.

Brainstorming sessions should be freewheeling, "think-outside-the-box" types of meetings. There are only a few rules. One is to never criticize or allow criticism of any ideas, because this can stifle the team's willingness to express ideas. The other is to focus on the quantity (and not the quality) of the ideas. The intent is to get people to open up and to generate as many potential causes as possible.

Mind Mapping

Mind mapping, developed by Tony Buzan in the early 1970s, is essentially a graphical outlining technique. What sets it apart from other outlining

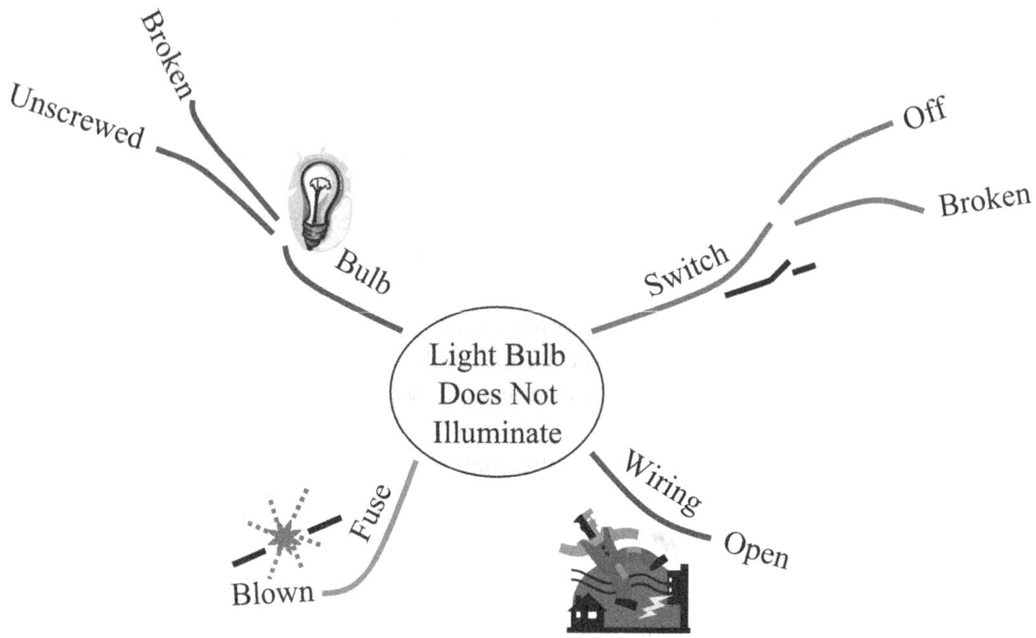

Fig. 4.1 Mind map for a light bulb failing to illuminate. Compare this image with Fig. 4.2 and with the fault-tree analysis prepared for the same failure in Fig. 5.4

techniques is that it is said to foster creativity through the use of images, curved lines, and colors. Mind-mapping images are free-flowing with an idea in the center, supporting concepts arranged around it, and lines connecting the supporting concepts to the center concept. In a problem-solving scenario, the central idea is the problem, with the outwardly radiating curves defining families of potential causes. Figure 4.1 shows a typical mind map applied to a light bulb failing to illuminate.

The intent of mind mapping is to defeat linear thinking. Linear thinking refers to thinking about things in a sequential, one-at-a-time manner. For example, these words are read sequentially, one at a time. Words are written one letter at a time; sentences are created one word at a time; and so on. Some believe that this linear, sequential, one-step-at-a-time approach inhibits creativity. The mind-mapping technique attempts to break out of this linear thinking by relying on images, curved lines, colors, and a graphical structure.

To implement the mind-mapping approach, the failure analysis team should use the following steps:

- Start in the center of the page. The approach involves taking a clean sheet of paper, preferably one that is unlined, and creating an image in the center of the page showing the problem to be solved. Due to the graphical nature of this approach, it is best to draw the mind map by hand (rather than using a computer). The earlier example is shown with a computer only for the sake of clarity.

- Draw curved lines, in different colors, radiating outward from the central image (i.e., the image depicting the problem). These curved, outward-radiating lines should show potential problem causes. The theory behind using curved lines is that straight lines may induce linear, restricted, creativity-inhibiting thinking. Curved lines are thought to foster creativity. The different colors are intended to further foster creativity.
- On each line, and for the central image, add single-word descriptions stating the problem and its potential causes. The concept behind using a single word is that it is less restrictive, allowing for greater creativity.
- On each line, add an image associated with the single word. The theory here is that people are used to thinking in images, and adding an image promotes greater creativity.
- Continue to develop the curved lines, showing subsets (or causes) of each hypothesized cause. As described earlier, use colors, curved lines, and single-word descriptions. Continue developing these subbranches until all possible failure causes are identified.

Ishikawa Diagrams

The Ishikawa diagram is another graphical technique used to identify potential failure causes. Kaoru Ishikawa, a quality manager at Kawasaki shipyards in the 1960s, developed the technique that bears his name. An Ishikawa diagram loosely resembles a fish skeleton (Fig. 4.2), and for this reason, Ishikawa diagrams are frequently referred to as fishbone diagrams. Sometimes, these graphical analyses are also called cause-effect diagrams.

As was the case with the mind-mapping technique, it is best to start the Ishikawa diagram with a large sheet of unlined paper.* The problem is expressed on the right side of the paper, with a horizontal line drawn across the page (Fig. 4.2). The failure analysis team can then add ribs (diagonal lines) to the chart to show failure-cause categories. For each of these categories, the failure analysis team can then add branches that show the underlying causes. It is a good idea to have failure analysis team members present as the Ishikawa diagram is created, because the team will have ideas to add bones to the fish's skeleton.

Some suggest using the four "M," "P," and "S" categories to assist in brainstorming Ishikawa diagram potential failure causes:

- *The four "M's":* These include methods, machines, materials, and manpower.
- *The four "P's":* These include place, procedure, people, and policies.

*Although Fig. 4.2 was computer generated, trying to use a computer during the brainstorming process can stifle creativity. When the failure analysis team has created the chart, it can be converted to a computer image using any of several software programs.

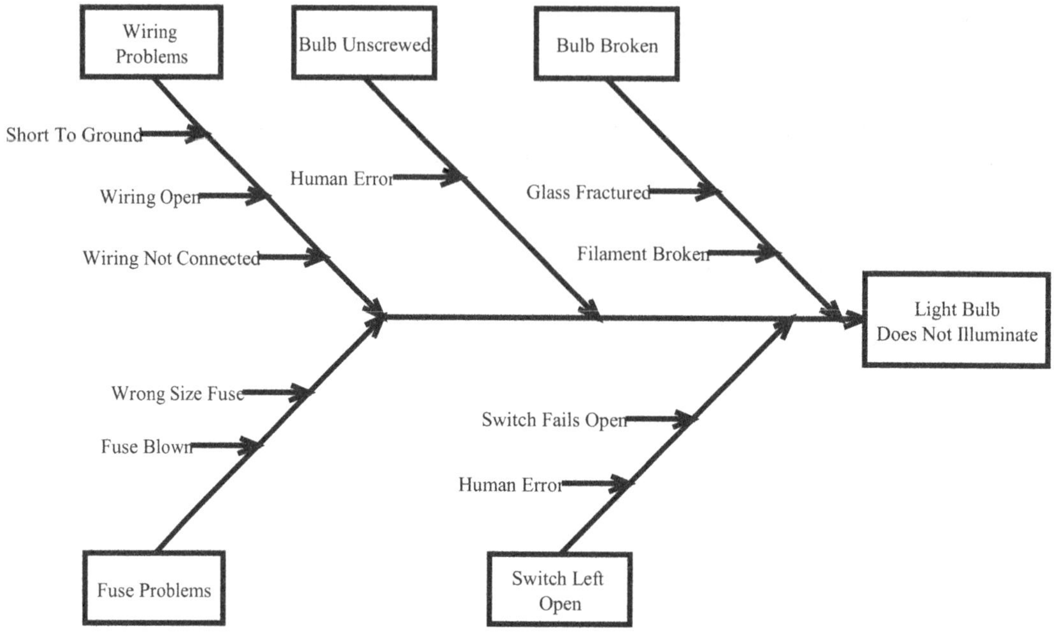

Fig. 4.2 Ishikawa diagram showing potential failure causes for a light bulb not illuminating. Ishikawa diagrams may help the failure analysis team identify more potential causes of failure than the mind-mapping technique.

- *The four "S's":* These include surroundings, suppliers, systems, and skills.

The aforementioned categories should be viewed as suggestions only. They can be helpful, but they should not constrain the failure analysis team members from examining other areas.

The "Five Whys" Technique

The "five whys" technique is another approach that some find useful for unearthing potential failure causes. Sakachi Toyoda originally developed the technique at Toyota Motor Company, and it became part of the famous Toyota Production System. It is also part of the Six Sigma approach to quality management.

The "five whys" approach is quite simple (some say trivially so). It is usually and perhaps best used to support brainstorming and Ishikawa diagram preparation. The following steps are involved in its use:

- State the problem.
- Ask why the problem occurred, and note the answer.
- Determine if the aforementioned answer defines the root cause of the problem.

- If the aforementioned answer does not identify the root cause of the problem, examine the answer and ask why it (i.e., the answer to the preceding "Why?") occurred.
- Continue to iterate the previous process until the group agrees that the root cause has emerged.

There is nothing magical about asking "Why?" five times. If the root cause is identified in fewer than five "Why's?", the analysis can be concluded. If the root cause has not emerged with five "Why's?", the process can be continued until the root cause has been identified.

The "five whys" approach has been criticized as being too superficial to be effective. Critics feel it can stop too early without unearthing the root cause of a failure, it does not lend itself to identifying all potential root causes, and it tends to focus on symptoms rather than underlying problems. Its principle value is that it can force the failure analysis team to go beyond simple answers and to dig deeper for underlying causes.

Flow Charting

Flow charting involves preparing a block diagram showing how the product or process works. (Figure 4.3 shows a flow chart for the light bulb used in the earlier examples). The failure analysis team starts with the action or activity that initiates the process or the product functioning and then shows succeeding blocks for each subsequent, sequential step. At each block in the flow chart, the failure analysis team should show required inputs and outputs. The steps involved in preparing a flow chart help the failure analysis team identify how the product or process works. If the team asks (at each block in the flow chart) if anything can happen to induce the failure, the flow chart becomes a useful tool for unearthing additional potential failure causes.

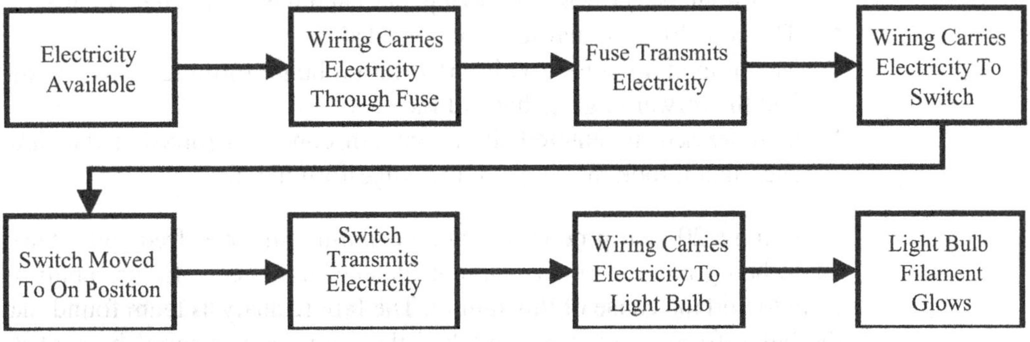

Fig. 4.3 Light bulb flow chart showing the sequence of steps required for the light bulb to illuminate

Summary

This chapter discussed five techniques for identifying potential failure causes:

- Brainstorming involves a group free-flow of ideas in which the ideas are not evaluated as they are generated. The failure analysis team should be encouraged to "think outside the box" in listing potential failure causes. In brainstorming, idea quantity is valued over idea quality. The ideas will be evaluated later.
- Mind mapping is a graphical technique emphasizing images, colors, curved lines, and single-word descriptions. The concept is to foster idea-generation creativity.
- Ishikawa (or fishbone) diagrams are another graphical technique. Ishikawa diagrams identify potential failure causes. Brainstorming, "five whys," and flow-charting techniques can be used to support Ishikawa diagram generation.
- The "five whys" technique states the problem, asks why it occurred, and then asks why the answer occurred. The theory is that repeatedly asking "Why?" for each succeeding answer will bring the failure analysis team to the root cause of the problem.
- Flow charting defines in a graphical manner how products and processes operate. Flow charting can help the failure analysis team better understand how the system operates.

While each of the aforementioned analysis techniques may be useful, none of them develop potential failure causes as rigorously as fault-tree analysis. The discussion of fault-tree analysis begins with Chapter 5.

Example for Group Discussion. The Sargent-Fletcher Company manufactured wing-mounted fuel tanks for various fighter jets using a bulge-forming process. The wing tanks had a sleek, curved forward profile called the ogive. The ogive-forming process included several steps:

- Aluminum sheet was cut into a pattern and then rolled into a cone.
- The cone longitudinal seam was welded.
- The aluminum cone was inserted into a bulge-forming die. (The die had the forward ogive shape of the tank.)
- Compressed air entered the aluminum cone and forced it outward, such that it took the shape of the bulge-forming die.

During a 20-year production run, the aluminum cones frequently burst while being bulge formed. Sargent-Fletcher assigned a failure analysis team to find the cause of this failure. The failure analysis team found that the bulge-forming process stretched the aluminum material beyond its yield point. Sargent-Fletcher solved the problem by partially bulge forming the aluminum cone (so it did not exceed the yield point), annealing the aluminum, and then bulge forming the partially formed cone to its final shape. After modifying the process in this manner, the ruptures disappeared.

Which of the failure-cause identification techniques discussed in this chapter would have been best for finding the cause of this problem?

If the cause of the ruptures had been a combination of factors, would any of the techniques discussed in this chapter have identified the cause?

Could Sargent-Fletcher have used any of the techniques discussed in this chapter before starting production to identify and control potential failure causes?

SELECTED REFERENCES

- T. Buzan and B. Buzan, *The Mind Map Book: How to Use Radiant Thinking to Maximize Your Brain's Untapped Potential,* The Penguin Group, 1996
- B. Clegg and P. Birch, *Instant Creativity: Simple Techniques to Ignite Innovation and Problem Solving,* Kagan Page Ltd., 2007
- K. Ishikawa, *Guide to Quality Control,* Asian Productivity Organization, 1982
- T. Pyzdek, *The Six Sigma Handbook,* McGraw-Hill, 2003

CHAPTER 5

Fault-Tree Analysis

IN THE 1950s AND 1960s, the United States developed its nuclear intercontinental ballistic missile (ICBM) program. Nuclear ICBMs were designed such that when launched, they could not be destroyed in flight by a self-destruct signal. Test missiles can be destroyed in flight, but operational missiles carrying nuclear payloads cannot be destroyed after launch. The Air Force weapon systems planners had a concern that potential enemies would be able to find these signals and command the missiles to self-destruct prior to reaching their targets. The Air Force required that when the missile launched, it could not be destroyed in flight.

Based on this requirement, another concern emerged: What features must the system have to prevent an inadvertent launch? Recall that in Chapter 3, a system failure was defined as a system not doing something when it is supposed to do it or doing something when it is not supposed to. An inadvertent launch of a nuclear missile certainly falls into this latter category.

A nuclear missile launch involves thousands of parts and numerous electrical signals, and the Air Force needed to make sure that the probability of these occurring in an uncommanded (or inadvertent) manner to launch a nuclear missile was very low. The Air Force realized that it had no analysis technique that could identify all events (or combinations of events) that could result in an inadvertent nuclear missile launch. Boeing, the Air Force's prime contractor for the nuclear missile system, shared this concern. The Air Force and Boeing needed an analysis technique that could analyze the missile, its launch system, the crew, and all other aspects of the weapon system to identify all potential causes of an inadvertent launch.

To address the aforementioned concern, the Air Force turned to Bell Laboratories to develop a technology that identified all single-point causes and combinations of causes that could result in a nuclear missile launch. In response to this requirement, Bell Laboratories (working with Boeing and the Air Force) developed the fault-tree analysis concept. Fault-tree analysis is a graphical technique that identifies all events and combinations of events that can produce an undesired event.

Fault-tree analysis starts with the top undesired event and then graphically develops all potential causes of that event. In a systems failure analysis situation, the top undesired event is the system failure. From a systems failure analysis perspective, the top undesired event of the fault tree is the same thing as the first step in the four-step problem-solving process. It is the problem definition.

When the Air Force, Boeing, and Bell Laboratories worked together to develop the fault-tree analysis technique, they hypothesized the top undesired event (inadvertent nuclear missile launch) and then used the fault tree to identify all possible causes. In this case, the event had not occurred. The Air Force, Boeing, and Bell Laboratories used fault-tree analysis to predict what could cause it. In systems failure analysis applications, the event (i.e., the failure) has occurred, and fault-tree analysis is used to identify events or combinations of events that could have caused it.

Several fault-tree analysis concepts are emphasized at this point:

- The fault-tree analysis is a technique that focuses on a single top undesired event. This is the thing that should not happen, and, as mentioned previously, it is the problem to be fixed or prevented. The system can fail to do what it is supposed to do; the system can do something it is intended to do but at the wrong time; or the system can do what it was designed to do but in some out-of-specification manner. The failure analysis team should define the fault-tree top undesired event to be consistent with the problem definition.
- The fault-tree analysis is a graphical analysis technique that can be computer generated or hand drawn. Computer-generated fault trees are orderly and present well, and they can speed fault-tree quantification (this is discussed in more detail in Chapter 7), but developing the charts by hand usually helps the failure analysis team identify more potential failure causes. Even on complex systems, it is preferable to prepare fault trees by hand and then convert them to a computer-generated format later.
- Fault-tree analysis does not evaluate all potential failure modes of all parts (as do other analytical techniques such as failure modes and effects analysis). The fault-tree analysis is only interested in failure causes, conditions, events, and combinations of these things that can induce the top undesired event. Other events or conditions that do not contribute to the top undesired event are irrelevant to the fault-tree analysis.
- In its quest to identify all potential causes of a top undesired event, the fault-tree analysis considers component failures, subassembly failures, normal conditions, human errors, and, as mentioned previously, combinations of these items. This ability to identify combinations of things that can induce the top undesired event is a major fault-tree analysis advantage. All other analysis techniques focus on specific component

failures in isolation. These other analysis techniques do not identify the effects of combinations of equipment failures, human errors, and normal conditions.

The last point is particularly relevant when seeking to identify failure causes in complex and sophisticated systems, which frequently have redundant features to prevent single-point failures from inducing a system failure.

Consider the General Dynamics F-16 fly-by-wire, quadruply-redundant example provided at the end of Chapter 3. As is frequently the case in large organizations, General Dynamics' F-16 flight control engineers created the circuit and control logic schematics that provided quadruple redundancy. These schematics then went to the electrical harness design group, and that group of designers routed the quadruply-redundant signals through common connector plugs throughout the aircraft. General Dynamics built the first several aircraft before an independent engineering group discovered (through fault-tree analysis) that a single connector opening anywhere on any harness carrying the flight control signals would result in loss of an aircraft. The connector plugs could disconnect through a connector plug failure or as the result of a simple human error to properly connect it.

In this case, fault-tree analysis found that a complex, quadruply-redundant flight control system on a modern combat aircraft could be defeated by any of several hundred previously unrecognized single-point failures, including human error (one of the most probable failure modes). It sounds incredible, but had the fault-tree analysis not uncovered this error, it is unlikely that it would have been found and corrected as early as it was in the F-16 development program.

Charting Symbols in Fault-Tree Analysis

Figure 5.1 shows the symbols used by the fault tree. There are three categories of symbols: events, gates, and transfer symbols. Events (as the name implies) are things that can happen and, either in isolation or in combination with other events, induce an undesired event. Gates show the relationship between events (in other words, if an event can induce an undesired condition by itself or if it must be combined with other events or conditions). Transfer symbols are used for transferring portions of the fault tree from one sheet to another.

Fault-tree analysis uses four different event symbols:

- *Command event:* The rectangle represents the command event. It is used for a condition that is induced by the events immediately below it (demonstrated subsequently).
- *Basic failure event:* The circle represents a basic failure event. These are typically component failures, such as a resistor failing open or a structural member fracturing.

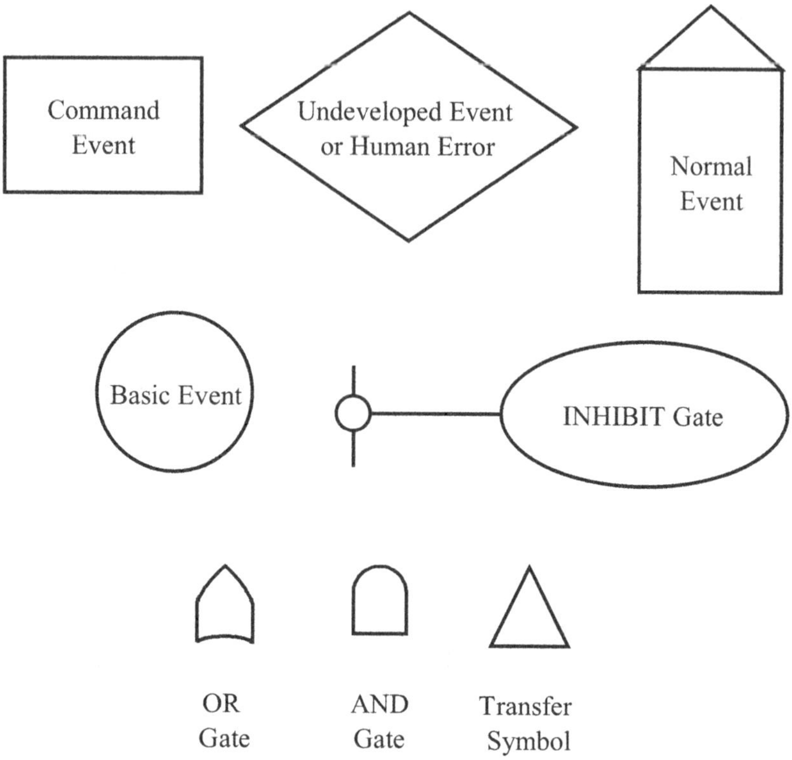

Fig. 5.1 Fault-tree analysis symbols representing different events and conditions and their relationships

- *Normal event:* The house represents a normally occurring event. (For example, if electrical power is normally present on a power line, a house symbol with the text "Power present" is used to represent this event.)
- *Human error or undeveloped event:* The diamond represents either a human error or an undeveloped event. A human error may be a pilot failing to extend the landing gear when landing an aircraft, a technician failing to properly adjust a variable resistor, or a missile launch crew member inadvertently depressing a self-destruct button on a control console. An undeveloped event is one that requires no further development. Events that are considered extremely unlikely are designated as undeveloped events to show that they have been considered and eliminated as possible failure causes.

As mentioned previously, gates show the relationships between events and conditions. There are three types of gates, as shown in Fig. 5.2 and explained as follows:

- *OR gates:* The OR gate means that if any of the events beneath it occurs, the event above it results.

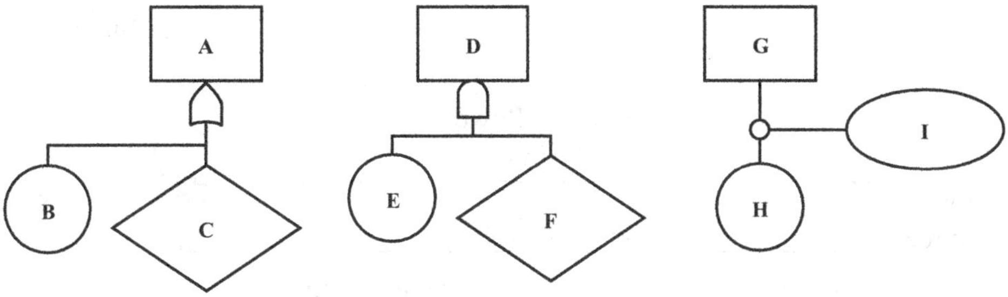

Fig. 5.2 Fault-tree gates. The OR gate (left) means the event above it will result if any event below it is present (if either B or C occurs, A will result). The AND gate (middle) requires all events below it to result in the event above it (E and F must occur for D to result). The INHIBIT gate (right) functions somewhat like an AND gate; H must occur, combined with the I condition in the ellipse, for G to occur.

- *AND gates:* The AND gate signifies that all events beneath it must occur simultaneously to result in the event above it.
- *INHIBIT gates:* The INHIBIT gate is a special form of the AND gate. The INHIBIT gate places a constraint on the event below the INHIBIT gate. The constraint is shown in an ellipse to the right of the INHIBIT gate. When the condition to the right of the INHIBIT gate is satisfied and the event below it occurs, the event above the INHIBIT gate occurs.

The transfer symbol is designated by a triangle. It shows a continuation of the fault-tree logic from one fault-tree sheet to another. Most fault trees are far too large to fit on a single sheet, so transfer symbols are used extensively. These are covered in more detail in Chapter 6.

The Light Bulb Fault-Tree Analysis

This section examines how all of the aforementioned comes together to generate a fault-tree analysis. Consider a simple system failure analysis for the same situation covered in Chapter 4—a system with a light bulb that screws into a socket, and the light bulb illuminates when the switch is moved to the "on" position. Figure 5.3 shows a schematic for this system. This system experiences a failure: When the switch is actuated, the light bulb fails to illuminate.

The first step in solving any problem is to define the problem. The problem here is that the light bulb does not illuminate. This becomes the top undesired event in the fault tree for this system failure, and Fig. 5.4 shows it in a command event (the rectangle symbol). Top undesired events are always shown in a command event symbol, because they are commanded to occur by the events below them.

The next step in performing a fault-tree analysis is to look for the causes at the point where the failure manifests itself or the immediately adjacent

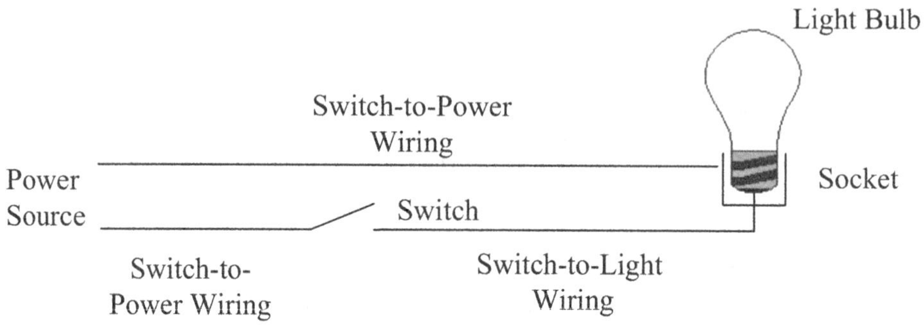

Fig. 5.3 Indicator light system. In this system failure, the light bulb failed to illuminate. The fault-tree analysis in Fig. 5.4 identifies all potential failure causes.

causes that can induce the command event. This idea of looking only for causes at the point where the failure manifests itself and the immediately adjacent inputs is a critically important concept. A common failure analysis shortcoming is to jump around in the system and start listing things such as a power loss in the building, a failed switch, or perhaps other upstream events. This is not desirable to do. The fault tree requires discipline. One must look for the causes at the point where the failure manifests itself and for immediately adjacent causes. In this case, it is necessary to locate the causes within and immediately adjacent to the light bulb.

An approach for doing this is to imagine oneself as the light bulb, screwed into the socket, and ask, "What can happen in me or right next to me to prevent me from illuminating?"

Considering only the causes within the light bulb and the immediately adjacent causes, the answers are:

- Open light bulb filament
- Fracture in the bulb glass
- Contaminated terminals in the socket
- Bulb not fully screwed into the socket
- No electrical energy from the socket

For now, the analysis does not go beyond the immediately adjacent causes (that comes later). The goal is only to identify what can happen in or right next to the event.

The next step in constructing the fault tree is to show these events immediately below the top undesired event and then to determine the appropriate symbol for each. Referring to Fig. 5.4 will help the reader to follow this discussion. Here is how these are developed:

- The open filament is a basic component failure (i.e., the filament failed open), so it is represented by the circle symbol.
- A fracture in the bulb glass is another basic component failure, so it also goes in a circle symbol.

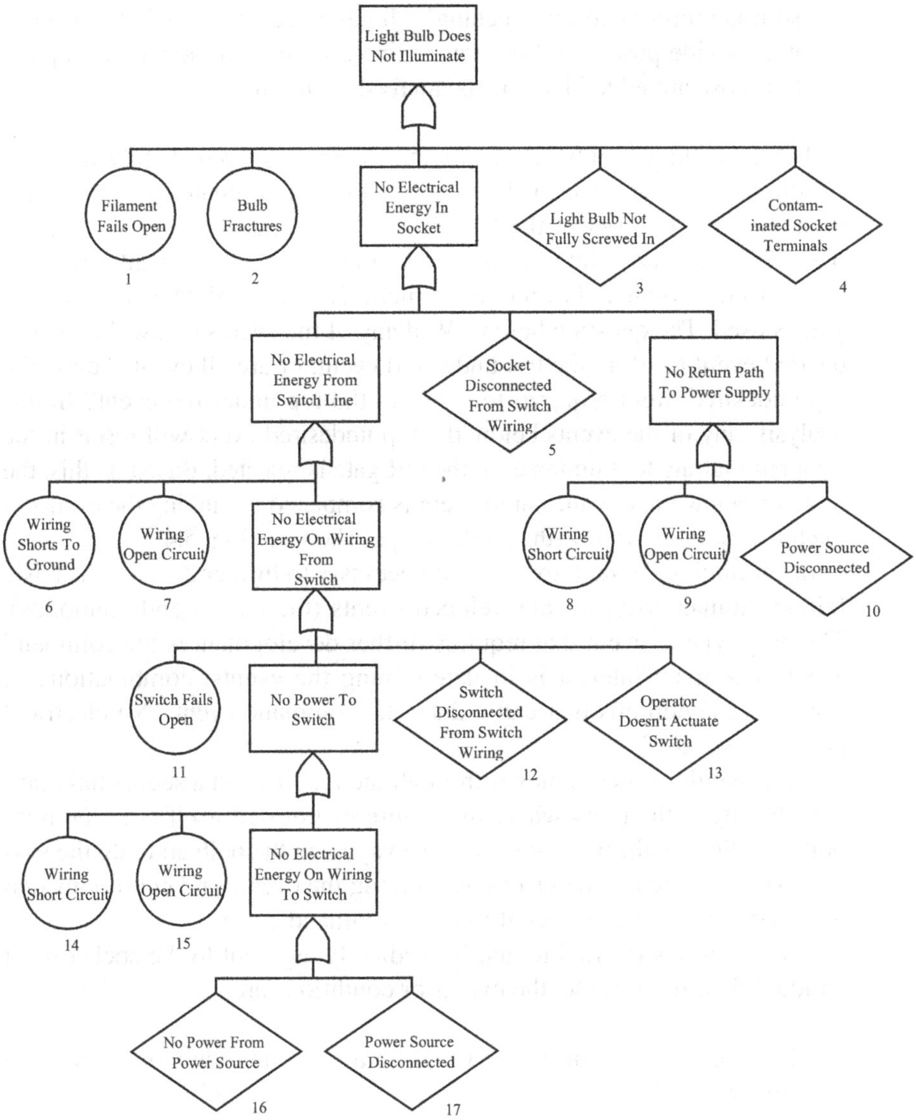

Fig. 5.4 Light bulb fault-tree analysis develops potential causes for the light bulb shown in Fig. 5.3 failing to illuminate.

- Contaminated terminals in the socket could be caused by a variety of conditions (a socket manufacturing error, dust or other contaminants entering the socket, etc.), but for the purposes of this analysis, these are not fully developed. The contaminated terminals condition is placed in an undeveloped event symbol (the diamond).
- Not fully screwing the bulb into the socket is a human error, so it is represented by a human error symbol (also a diamond).
- No energy from the socket is a condition that is commanded to occur if other events occur elsewhere in the system. This is a command event,

so it is represented by a rectangle. It can occur if the switch line does not provide power or if the return line does not provide a return path, but those are added later as the analysis continues.

The preceding events are all of the internal or immediately adjacent conditions that can cause the light bulb to fail to illuminate, and this nearly completes the first tier of the fault tree for this undesired event. To complete this tier, these internal and immediately adjacent events must be linked to the command event above them. Either an AND gate or an OR gate is used. The question here is: Will any of the events below the top undesired event result in the top undesired event, or are all events below the top undesired event required to result in the top undesired event? In this analysis, any of the events below the top undesired event will result in the light bulb failing to illuminate, so the OR gate is selected. Based on this, the first tier below the top undesired event is completed by linking the events to the top undesired event with an OR gate, as shown in Fig. 5.4.

In continuing the fault tree, it is unnecessary to further develop the basic failure, human error, or undeveloped events (the circles and diamonds). The only type of event that requires further development is the command event. The main interest is in determining the events, combinations of event, and conditions that can induce the command event "No electrical energy in socket."

Note that this process moves through the fault tree in a sequential manner, starting at the point where the failure manifested itself (i.e., the light bulb not illuminating) and navigating away from the bulb and into the system. This is done in small steps, examining the events and conditions that are in or immediately adjacent to each command event.

All conditions internal to and immediately adjacent to the socket must be identified. In this case, the events or conditions are:

- The socket can be disconnected from the wiring (which is shown as a human error).
- The switch wiring can have no power delivered to it (which is shown as a command event).
- The return line could have no return path (which is also shown as a command event).

These three events are shown on the third tier of the fault tree. Any of them can induce the command event above, so it is appropriate to use an OR gate.

The next step is to develop the two command events on the third tier (no power from the switch wiring and no path to ground). The "No electrical energy from switch line" portion of the fault tree is developed first. The "No return path to power supply" fault-tree branch is developed later. Note that when the "No electrical energy from switch line" portion of the fault tree is done, the attention is focused exclusively on this fault-tree branch, and the process continues to navigate in a point-to-point manner. Because the objective is to look for events and conditions that can prevent power from the switch line reaching the socket, the following are shown:

- Wiring open circuit (as a basic failure event)
- Wiring short circuit (as a basic failure event)
- No power going into the switch line (as a command event)

The last item (no power going into the switch line) is shown as a command event because it is developed further. Any of the preceding events can induce "No electrical energy from switch line," so an OR gate is used above these events.

The next command event to develop is "No electrical energy on wiring from switch." The events and conditions that can induce this command event include:

- Switch fails open (a basic failure event)
- Switch is disconnected from the wiring (a human error event)
- Operator does not actuate the switch (a human error event)
- No power to the switch (a command event to be developed further)

Again, any of the preceding events can induce "No electrical energy on wiring from switch," so an OR gate is used above these events.

The command event on this tier, "No power to switch," can be induced by:

- Wiring open circuit (as a basic failure event)
- Wiring short circuit (as a basic failure event)
- No electrical energy on the wiring to the switch (a command event)

Because any of the preceding can induce the command event above them, an OR gate is used again.

At this point, it is necessary to develop the causes for no energy going into the wiring from the power supply, which include:

- No power from the power supply (an undeveloped event)
- Power supply disconnected (an undeveloped event)

These are shown as undeveloped events, because these events are not developed further. This may be because the power supply is a purchased item, or there may be interest in only analyzing potential failure causes upstream of the power supply. Again, the use of an OR gate is appropriate here, because either event can induce the command event above.

There is still one more fault-tree analysis branch to develop. Recall that the second tier of the fault tree showed another command event, "No return path to power supply." Potential causes of this command event include:

- Wiring short circuit (a basic failure event)
- Wiring open circuit (a basic failure event)
- Power source disconnected (an undeveloped event)

Note that the last event in the preceding list is the same event used in another branch of the fault tree. This same event could also be shown going into the two different gates on both sides of the fault tree. Either approach is acceptable. Once again, any of the aforementioned conditions can induce the command event above, so an OR gate is used.

Before leaving the fault tree, there is one more task. A unique number must be assigned to each of the basic events, human errors, normal events, and undeveloped events. These are used for tracking purposes, as explained later in Chapter 8.

At this point, the fault-tree logic for this simple example is complete because all of the events in the fault-tree branches are basic failure events, undeveloped events, human errors, or normal events. If there were any command events in the fault tree that had not been developed, the fault tree would not be complete.

To review, with the data available, the fault tree started with a definition of the failure (which became the top undesired event in the fault tree). All potential causes of the failure were systematically developed. It is important to note that the fault-tree logic development started at the point where the failure manifested itself. (In this case, the light bulb failed to illuminate, so the process started at the light bulb). The fault tree then progressed through the system in a disciplined manner, moving point to point and following the system design. Systematically working from one point to the next when constructing a fault tree forces the analyst to consider each part of the system and all system interfaces (such as where the switch interfaces with a human being). This is a key feature to successful fault-tree construction that allows the fault-tree analysis to identify all potential failure causes.

Summary

Fault-tree analysis is a graphical technique that allows failure analysis teams to identify all potential causes of an undesired event. In the fault-tree analysis context, the undesired event is the problem definition; that is, it is the problem the failure analysis team is going to solve.

When preparing a fault-tree analysis, the failure analysis team should mentally place themselves at the point where the failure manifested itself and ask, "What is in me or immediately adjacent to me that can induce this condition?" After answering this question, the events and conditions that can induce the failure are further developed, navigating the through the system in a point-to-point manner.

If any of the hypothesized causes of a command event could induce the condition, an OR gate should be used. If all of the hypothesized causes going into a command event are required, an AND gate should be used.

Example for Group Discussion. In the Chapter 4 examples, potential light bulb illumination failure causes included the fuse. The fuse was not included when preparing the fault-tree analysis in this chapter. Why? If fuse-related causes were to be included in the fault-tree analysis, where would these be shown?

Why did the fault-tree analysis in this chapter identify more potential failure causes than did the techniques covered in Chapter 4?

Fault-tree analysis is a technique that identifies all potential failure causes. This includes many causes beyond those that caused the actual failure being investigated. Should the failure analysis team implement corrective actions only for those causes proven to have induced the failure; should the team implement corrective actions for all potential failure causes; or should the team attempt to select a subset of the potential failure causes and implement corrective actions only for these?

SELECTED REFERENCES

- *Fault Tree Construction Guide,* Armament Development Test Center, U.S. Air Force, 1974
- W.F. Larsen, *Fault Tree Analysis,* Ammunition Reliability Division, Picatinny Arsenal, U.S. Army, 1968
- L. Stewart and J.H. Berk, Ordnance System Failure Analysis, *Proceedings of the Annual Reliability and Maintainability Symposium,* 1987

CHAPTER **6**

Fault-Tree Analysis Special Topics

CHAPTER 5 PROVIDED a fault-tree analysis foundation. This chapter covers special fault-tree analysis topics, including the use of:

- Transfer symbols
- INHIBIT gates
- Sequential AND gates
- State-of-the-part versus state-of-the-system considerations
- Use of the fault tree to identify redundancy-defeating failures

Transfer Symbols

A fault-tree analysis for a complex system failure analysis can include hundreds (sometimes thousands) of events and conditions. Not all of these will fit on a single sheet, so fault-tree analysis uses transfer symbols to transfer logic (the fault-tree graphics) from one sheet to another.

The transfer symbol is a triangle. Each fault-tree analysis sheet (other than the first sheet) has a transfer symbol out of the command event at the top of the sheet (Fig. 6.1). This is called a transfer out because the fault-tree logic is transferring out of this sheet to another sheet. The transfer symbol has a transfer designation number inside the triangle. The transfer symbol also has text next to the triangle indicating which fault-tree analysis sheet the logic transfers to or from.

As the fault-tree logic is developed on each fault-tree analysis sheet, the logic includes basic events, undeveloped events, normal events, and command events. There may be room on the fault-tree analysis sheet to develop the logic beneath lower-tier command events, but, in most cases, more space is needed to continue the development.* To extend the logic to

*The light bulb fault-tree analysis included in Chapter 5 only required a single fault-tree analysis sheet because it is a very simple system.

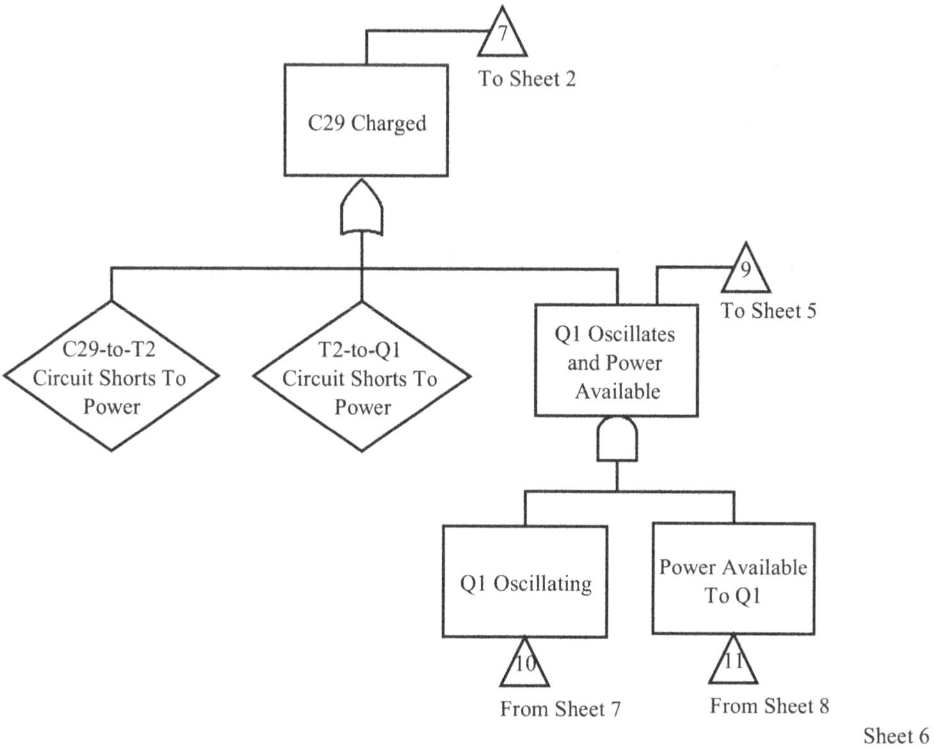

Fig. 6.1 Fault-tree transfer symbols. This is a sheet (sheet 6) taken from a fault-tree analysis developed for a circuit card assembly failure. Transfer 7 shows that the sheet 6 fault-tree logic goes into sheet 2 of the fault-tree analysis. Transfer 9 shows that the event "Q1 oscillates and power available" goes into sheet 5 of the fault-tree analysis. Transfer 10 shows that the logic for "Q1 oscillating" comes from sheet 7 of the fault-tree analysis, and transfer 11 shows that the logic for "Power available to Q1" comes from sheet 8 of the fault-tree analysis.

another fault-tree analysis sheet, a transfer symbol is used beneath the command event whose logic is developed on a subsequent sheet (Fig. 6.1). These transfer symbols also receive transfer designation numbers and text showing the fault-tree analysis sheet where the logic is continued.

When transferring from one fault-tree analysis sheet to another, the command event and the text it contains are duplicated on both sheets. On one fault-tree analysis sheet (the sheet showing that the logic beneath the command event is continued on another sheet), the command event has the transfer symbol below it and text showing the sheet where the logic is continued (e.g., "From sheet 7"). On the fault-tree analysis sheet where the logic is continued, the identical command event is at the top of the sheet, with the same transfer symbol out and text showing where the logic goes (e.g., "To sheet 2").

During fault-tree analysis preparation, the fault tree analyst may recognize that portions of the fault-tree logic for particular system segments may apply to other fault-tree areas. In this case, the fault-tree analyst can show the same command event transferring to other fault-tree areas. This

Table 6.1 Fault-tree analysis transfer log showing all fault-tree transfers, allowing for easy reference and tracking of fault-tree logic

Transfer number	Command event	From sheet	To sheet
7	C29 charged	6	2
9	Q1 oscillates and power available	6	5
10	Q1 oscillating	7	6
11	Power available to Q1	8	6

is done in the same manner as a command event transfer at the top of the sheet. Figure 6.1 also shows this concept (see transfer 9).

As the fault tree grows, so will the number of transfer symbols. This can become confusing, so it is a good idea to initiate and maintain a fault-tree transfer log as the fault-tree analysis progresses. The fault-tree transfer log shows each command event that has an associated transfer, the command event description, the transfer symbol number, and the sheets between which the logic transfers. Table 6.1 shows the concept.

Transfer symbols are only associated with command events. Basic, undeveloped, and normal events do not transfer by themselves to other fault-tree sheets.

Inhibit Gates

In Chapter 5, INHIBIT gates were mentioned as a special form of the AND gate. The INHIBIT gate takes the event below it and adjusts or conditions it to result in the condition above, as shown in Fig. 6.2.

Most of the time, the condition described in the INHIBIT ellipse can be tied to a probability. For example, suppose the loss of an aircraft hydraulic system is being examined. If a fault-tree analysis is created for loss of hydraulic power, at some point those portions of the hydraulic lines that run to the aircraft landing gear must be considered. One of the potential failure causes for losing hydraulic power is an aircraft tire blowout in the wheel well, with aircraft tire fragments penetrating any of the hydraulic lines in the wheel well. If the aircraft tire exploding in the wheel well is visualized, it can be understood that in some cases the aircraft tire fragments will hit and rupture a hydraulic line, and in other cases they will not. If the command event is "Aircraft tire blowout ruptures hydraulic line," only a portion of aircraft tire blowouts will actually break a hydraulic line. That situation can be modeled as shown in Fig. 6.2.

Upon further consideration, it is recognized that two INHIBIT gates are needed to more accurately model the situation described previously. One is that a tire fragment strikes a wheel well hydraulic line with sufficient force to rupture it, and the other is that a wheel well hydraulic line is struck at all. Figure 6.3 shows these stacked INHIBIT events. Each INHIBIT gate ellipse assigns a probably to the event beneath it. Given that the tire explodes, there is a probability that a tire fragment will strike one of the wheel well

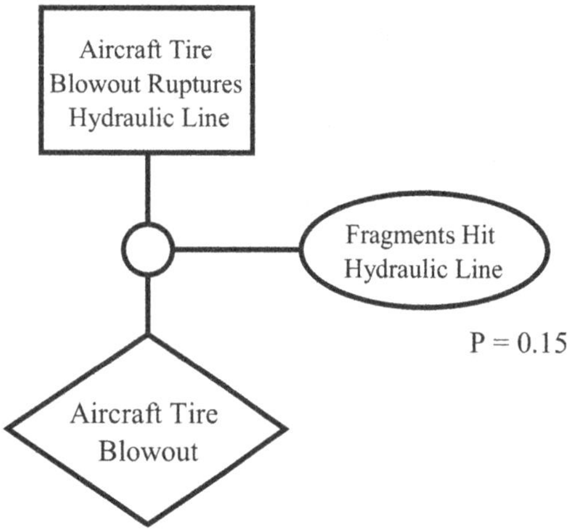

Fig. 6.2 INHIBIT gate. The undeveloped event "Aircraft tire blowout" combined with the INHIBIT condition "Fragments hit hydraulic line" cause the command event "Aircraft tire blowout ruptures hydraulic line."

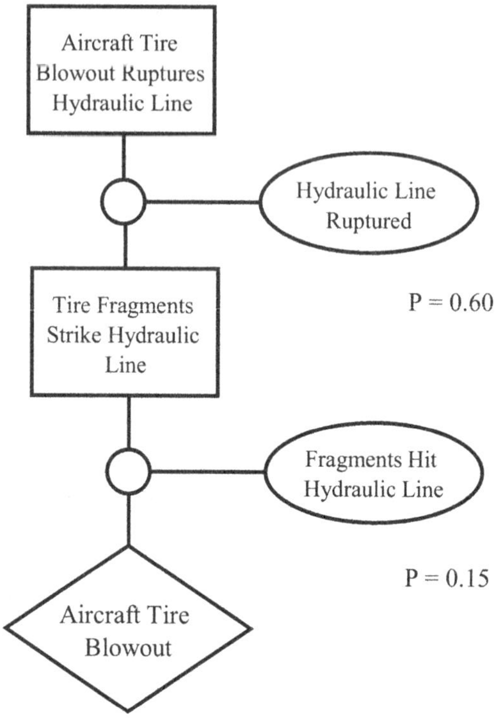

Fig. 6.3 Stacked INHIBIT gates. This stacked INHIBIT gate situation more accurately simulates the events and conditions required for a hydraulic line rupture if a tire blowout occurs in the aircraft wheel well.

hydraulic lines, and given that one of the wheel well hydraulic lines is hit by a tire fragment, there is a probability that it will be struck with sufficient force to rupture the line. The stacked INHIBIT gates in Fig. 6.3 show this. Estimating the probability of these inhibiting conditions is a topic covered in Chapter 7.

INHIBIT gates always go into command events. Any type of event can go into the INHIBIT gate (normal events, undeveloped events, human errors, command events, and basic failure events).

Sequential AND Gates

Ordinarily, an AND gate below a command event means that if all of the conditions going into the AND gate are satisfied, the command event above the AND gate occurs. In Fig. 6.1, for example, if both of the command events below the AND gate occur, the command event above it occurs.

However, in some special cases, not only must all of the AND gate inputs occur, but they must occur in particular sequence for the command event above the AND gate to occur. Figure 6.4 shows how this situation is modeled using fault-tree analysis.

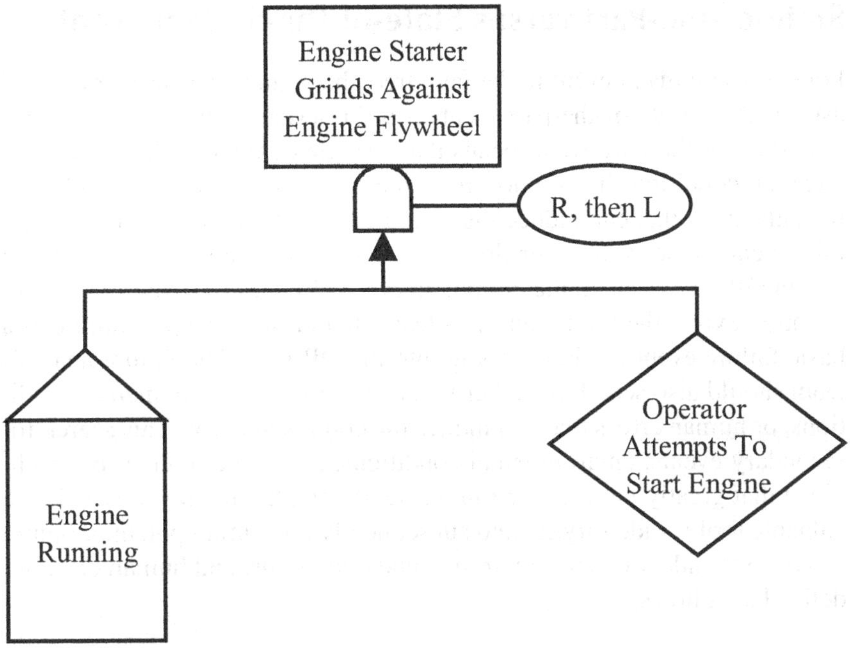

Fig. 6.4 Sequential AND gate. Both AND gate inputs must be present, and they must occur in the sequence specified.

An example most people can relate to is attempting to start an automobile engine when it is already running. Suppose the command event being modeled in a fault tree is "Engine starter grinds against engine flywheel." To satisfy that condition, the engine must be running, and the operator must attempt to start the engine. Based on the aforementioned, the situation is modeled with an AND gate, as shown in Fig. 6.4. However, in this case, the sequence is significant. In normal operation, the operator attempts to start the engine, and then the engine starts. For a brief period, both conditions are satisfied, yet in this situation, the starter will not grind against the engine flywheel. If the engine is already running when the operator attempts to start the engine, the starter will grind against the flywheel.

In modeling this situation, the sequence of events going in the AND gate is important, and the command event above the AND is only satisfied if the events occur in a particular sequence. This is designated with text to the right of the AND gate stating "Right, then left," indicating the special sequence.

There is another significant nuance in the example shown in Fig. 6.4. Notice that "Operator attempts to start engine" is shown as a human error. It could also be shown as a normal event, but the human error symbol was chosen because, in this case, it is a mistake to attempt to start the engine when it is already running.

State-of-the-Part versus State-of-the-System Events

For every command event in the fault tree, the failure analysis team should ask, "Is this a state-of-the-part event, or is this a state-of-the-system event?"

A state-of-the-part event means that a single component failure can induce the condition. It does not mean that the single component failure is the only event that can induce the condition, but it does mean that a single failure can do so. Whenever this occurs, the failure analysis team should use an OR gate beneath the command event. Whenever a state-of-the-part situation exists, the failure analysis team should show the part failure as a basic failure event (a circle) going into the OR gate. The failure analysis team should also search for other secondary events, environmental conditions, or human errors that can induce the command event. This search for secondary events, environmental conditions, and operator error is a technique that greatly increases the universe of potential failure causes. It is a valuable tool for identifying and subsequently preventing potential failure causes. Secondary events, environmental conditions, and human error are defined as follows:

- *Secondary events:* A secondary event is another command event. In other words, can the command event above the OR gate be induced by another command event in addition to the part failure? For example, if a command event shows inadvertent electrical energy presence on a wire, a part failure could be the wire insulation allowing it to short circuit to power (a basic failure event). If energy can also be provided to the wire by a switch, that is a command event, or a secondary event. This is what is meant by a secondary condition; it is a command event below the OR gate going into the command event.
- *Environmental conditions:* An environmental condition is a special environment that can induce the command event above the OR gate. Suppose the failure analysis team identifies a situation in a fault tree in which a command event indicates "No power present." One failure cause could be that the wire carrying power fails open (a basic failure event). Because this condition can be induced by a single part failure, it is a state-of-the-part situation, and because of that, the failure analysis team should look for environmental causes that can induce the condition. Either shock or vibration (of sufficient magnitude) can cause the wire to fail open. These events would most likely be modeled as either normal events (if they are normally present) going into an INHIBIT gate (indicating the magnitude was sufficient) or undeveloped events if the failure analysis team wishes to show that such events are hypothetically possible.
- *Human error:* A human error means that a human being does not do something he or she is supposed to do, or they do something they are not supposed to do. In the preceding example, the failure analysis

team, recognizing that this is a state-of-the-part situation, must look for human error. Another reason that power may not be present where it is needed is that the wire is not connected, either as the result of a manufacturing or maintenance error or because it is intentionally disconnected (perhaps as an act of sabotage).

Whenever state-of-the-part situations exist, the failure analysis team should search for secondary events, environmental conditions, and human error. These may not always be present, but the failure analysis team should always ask if such conditions are additional failure causes. Asking this question at each state-of-the-part situation frequently unearths previously unidentified failure modes. This is a powerful tool.

When asking if a command event can be induced by a single part failure, the answer is sometimes no. This means two or more conditions must be present to attain the command event. When this occurs, the command event is a state-of-the-system situation, and it should have an AND gate beneath it. The question at each command event pertains specifically to that event. There may be single-point failures (specific component failures) downstream that can induce the failure, but if they are not immediately adjacent to the command event under consideration, they should not be included in the fault tree at this point.

To summarize, when preparing a fault-tree analysis, the failure analysis team should:

- Start at the point where the failure manifests itself
- Work through the system using a point-to-point navigation of the system design
- Identify at each point those conditions that can induce the condition
- Determine if the condition at each point is a state-of-the-part versus state-of-the-system condition
- For state-of-the-part situations, use an OR gate and search for secondary, environmental, and human errors
- For state of the system situations, use an AND gate

The preceding approach enormously magnifies the universe of potential failure causes. While this increases the failure analysis team's workload, it is necessary work to identify and evaluate all potential failure causes.

Redundancy Defeaters

Sophisticated system designs often contain provisions for redundancy, which means incorporating additional design features such that if a single failure occurs, the total system will not fail. Chapter 3 described the quadruply-redundant circuit paths for the F-16 fly-by-wire system. Although the F-16 designers incorporated four independent circuit paths to carry each signal (with the thought that any three out of the four signals could be lost and the pilot could still fly the aircraft), any of several hundred connector plugs disconnecting in the original design would have resulted in loss of an aircraft. That undesirable failure mode was a redundancy defeater. A single failure would simultaneously eliminate all four levels of redundancy. The failure mode was subtle until it was discovered through fault-tree analysis (then it became glaringly and embarrassingly obvious).

In Chicago many years ago, another redundancy-defeating failure killed hundreds of people in an airplane crash. A McDonnell Douglas DC-10 jumbo jet lost an engine and crashed shortly after takeoff at Chicago's O'Hare field with a full passenger load. The DC-10 had been designed to fly if any one of its three engines lost power, and it had triply-redundant hydraulic systems. How was it, then, that the loss of a single engine introduced such a tragic event? The accident investigation revealed that an engine was incorrectly installed. This allowed the engine to literally tear free from the wing. When that happened, it tore the hydraulic lines to all three of the aircraft redundant hydraulic systems, resulting in a complete loss of hydraulic power. The pilot could not control the aircraft. Loss of the engine (a single-point failure) cascaded into a failure of multiple redundant systems. A single-point failure defeated all three levels of redundancy.

Fault-tree analysis is a useful tool for identifying redundancy defeaters for several reasons. First, it identifies all events, conditions, and combinations of events and conditions that can induce a failure. That means it identifies all of the combinations required to defeat redundant design features. Ideally, for redundant design features, the fault-tree analysis shows that the system requires multiple failures to defeat the design redundancies. In some cases, the fault tree shows that single-point failures can defeat all of the redundancies.

When the fault tree is complete and has identified these combinations of events and conditions, it reveals a complete set of conditions that can induce the top undesired event. On simple systems with no redundant design features, the list of failure causes includes many single-point failures or human errors. On more complex systems incorporating redundancy to prevent or at least greatly reduce the probability of undesired events occurring, the failure analysis team should ask these questions:

- What is the probability of these combined independent events and conditions occurring? (Chapter 7 addresses this issue.)
- Are the conditions identified by the fault tree independent, or are there single-point failures or conditions that can defeat the redundancy?

There are a couple of approaches to the second question. The first is to identify each of the events in the redundant (and presumably independent) fault-tree branches (with each fault-tree branch representing one of the redundant design features). If there are any events or conditions that are duplicated in each redundant branch, those branches are not independent, and they are therefore not redundant. Consider the F-16 fly-by-wire connector plug example discussed previously. Those connector plugs appeared in each independent branch of the fly-by-wire fault tree. A review of the events in each fault-tree branch showed that the same events (connector plugs disconnecting) appeared in all branches. When that condition surfaced, the General Dynamics designers realized the quadruply-redundant signal paths could be defeated by any of several hundred single-point failures.

The second approach involves taking a step back from the fault tree and contemplating all of the events and conditions associated with the independence of the redundant features. The failure analysis team must consider these questions:

- Are there any undeveloped events left in the branches of the fault tree that, if further developed, may reveal a common cause?
- Are there any environmental conditions that could simultaneously induce events or conditions in the independent fault-tree branches?
- Are there any human errors that, if present in one of the independent fault-tree branches, are likely to also be present in other independent fault-tree branches? In other words, is a human being likely to make the same mistake in several areas, which, when combined, can induce the undesired condition?
- Are there any failures that can occur in an independent fault-tree branch and remain undetected for some period such that, in effect, only one additional independent failure is required in another fault-tree branch to induce the undesired condition?

Summary

Most fault-tree analyses are too large to fit on a single sheet, so transfer symbols are used to show the continuation of fault-tree logic from one sheet to another. Transfer symbols on top of a command event show to which sheet the fault-tree logic transfers. Transfer symbols at the bottom of a command event show which sheets continue the fault-tree logic. Transfer logs help to keep track of the transfers in a fault-tree analysis.

INHIBIT gates modify (or condition) an event going into the command event above the INHIBIT gate. INHIBIT gates behave somewhat like AND gates in that the event below the INHIBIT gate is combined with the inhibiting condition to result in the command event above the INHIBIT gate.

Sequential AND gates require the presence of all conditions going into the AND gate in a specified sequence. Sequential AND gates are designed by text along the AND gate showing the required sequence.

OR gates should be used whenever state-of-the-part situations exist. For each state-of-the-part situation, the failure analysis team should assess if secondary events, environmental conditions, or human error could induce the command event above the OR gate. All three of these items may not always be present, but the failure analysis team should always consider if such conditions or events may occur. This approach greatly increases the universe of potential failure causes.

One of the fault-tree strengths is that it can identify redundancy-defeating failures. In many cases, the system design is truly redundant. In such situations, the fault tree identifies all combinations of multiple events and conditions required to defeat the redundancy. In other cases, single-point failure events or conditions can defeat multiple redundant design features. The fault-tree analysis identifies these situations.

Example for Group Discussion. As described previously, fault-tree analyses reveal if single-point failures can defeat redundant design features. This requires that the failure analysis team review different fault-tree branches to determine if the same groups of events or single event appear in different branches. It may be difficult to remember events appearing in different parts of large fault trees. What can the failure analysis team do to automate the search for similar or identical events in different fault-tree branches?

SELECTED REFERENCES

- *Fault Tree Construction Guide,* Armament Development Test Center, U.S. Air Force, 1974
- W.F. Larsen, *Fault Tree Analysis,* Ammunition Reliability Division, Picatinny Arsenal, U.S. Army, 1968

CHAPTER 7

Fault-Tree Analysis Quantification

FAULT-TREE ANALYSIS event probabilities and ranking of failure causes based on these probabilities is addressed in this chapter. Failure rates, failure-rate sources, probability determinations, mean times between failure, and related topics are also discussed.

When Bell Laboratories, the Air Force, and Boeing originally developed the fault-tree analysis technique, their objectives included quantifying the probability of an inadvertent launch. The fault-tree analysis has this capability. On many engineering development efforts (especially for military systems and nuclear power plants), the system must meet a quantified safety requirement related to the probability of occurrence. Fault-tree analysis is used for making this determination. This capability is also useful in the failure analysis environment, because it allows the failure analysis team to assess which hypothesized failure causes are more likely to occur.

Probability, Failure Rate, and Mean Time Between Failures

Probability is a dimensionless number that quantifies the likelihood something is going to occur. It ranges from 0 to 1 (sometimes also referred to as 0 to 100%). If the probability of occurrence is 0, it means it will never occur. If the probability of occurrence is 1, it means it will always occur. If the probability of something is 0.5, it means it will occur 50% of the time. Tossing a coin is frequently used to demonstrate this concept. Most people can readily accept that if a coin is tossed often enough, it will come up heads approximately 50% of the time and tails the other 50%.

The mean time between failures (MTBF) is the total number of hours of operation divided by the number of failures. Assume a particular type of resistor has experienced a total of 850,000 hours of service, and during this time it has had 29 failures. This means the average time between failures is:

$$\text{MTBF} = 850{,}000\ \text{h}/29\ \text{failures} = 29{,}310\ \text{h between failures}$$

Many component manufacturers (especially electronic component manufacturers) publish MTBF values for their components. Sometimes, subassembly manufacturers can similarly provide MTBF values for their products.

A failure rate (usually represented by the Greek letter λ) is also a measure of how often something is going to fail. Failure rate is the inverse of the MTBF. It represents the average number of failures that occur in one hour. For the component considered earlier, the failure rate is:

$$\lambda = 1/\text{MTBF} = 1/29{,}310 = 3.412 \times 10^{-5}\ \text{failures per hour}$$

Due to their very low values, failure rates are usually expressed in scientific notation.

If the operating time and failure rate are known, this information can be used to calculate the probability of success and the probability of failure. The probability of success (P_S) is determined with the following equation:

$$P_S = e^{-\lambda t}$$

where e is the natural logarithm base, 2.718282; λ is the failure rate, expressed in failures per operating hour; and t is the operating time, expressed in hours.

The probability of the part failing (P_F) during the same time period is 1 minus the probability of it working:

$$P_F = 1 - P_S$$

For example, if another resistor has a failure rate of 2.2×10^{-6} failures per hour and is operated for 5.5 h, its probability of working during that time period is:

$$P_S = e^{-(2.2\times 10^{-6})(5.5)} = 0.999988$$

The probability of the resistor failing during that same time period is:

$$P_F = 1 - P_S = 1 - e^{-\lambda t} = 1 - 0.999988 = 0.000012 = 1.2 \times 10^{-5}$$

The calculated probability of failure is a small number. It is just over 1 chance in 10,000 that the part will fail during the 5.5 hours it is operated. This seems small, but this is the probability of just that one resistor failing. If there are approximately 10,000 parts in the system and each one has approximately a 1-in-10,000 probability of failing, the system is likely to experience a failure of at least one part during its 5.5 hours of operation.

Fault-Tree Analysis Probability Determination

As mentioned previously, the failure analysis team can quantify the events in a fault tree. This is done to assess which of the failure causes is most likely to occur and to identify which of the failure causes requires corrective actions to prevent future occurrences. As explained later, this also determines the probabilities of different fault-tree branches occurring and the probability of the top undesired event occurring.

In fault-tree analysis quantification, the following practices are observed:

- Human errors are shown with probabilities (not failure rates). Probabilities for human errors typically range from 1×10^{-5} to 3×10^{-1}.
- Normal events are shown with probabilities (not failure rates). Probabilities for normal events can range from relatively low values all the way up to 1, depending on the expected frequency of occurrence.
- Basic failure event probabilities are calculated using component failure rates and operating times, as defined earlier.
- Undeveloped event probabilities are determined either by probabilities or by using failure rates and operating times.
- Inhibiting conditions (used with INHIBIT gates) are assigned probabilities based on the expected frequency of occurrence.
- Command event probabilities are determined by using equations associated with the gate type going into the command event (as explained subsequently) and the probabilities of the events going into these gates.

Each of these concepts is subsequently explained in more detail.

Human Error Probabilities. The failure analysis team can determine human error rates based on experiential factors, or the team can use published human error rate sources. Kirwan (Ref 1) provides a table defining human error probabilities under varying conditions (Table 7.1).

Table 7.1 Human error probabilities vary as a function of stress level and human-machine interface design

Operating condition	Human error probability
High stress level	0.3–0.3
Emergency situation	0.1
Routine operation requiring extra care	0.01
Simple routine operation	0.001
Switch selection	0.001
Best case for single operator	0.0001
Best case for team of operators	0.00001

One could very much question the lower probabilities shown in Table 7.1 (because making only one mistake in 10,000 or 100,000 operations when depending on human beings to do things correctly seems to be expecting too much). The author's experience indicates that under best-case conditions, human error rates are in the range of 1×10^{-2} to 1×10^{-3}. Under high-stress, emergency conditions, human error rates could be as high as 1. The important thing to recognize is that human error rates are relatively high (certainly much higher than component failure rates).

Normal Event Probabilities. The failure analysis team can determine normal event probabilities by assessing the likelihood of the event occurring. For example, suppose one wishes to assess the probability of an on-off switch being in the ON position when it is supposed to be. If it is assumed the system is already operating, the probability of the normal event "Power switch in ON position" is 1.

Suppose, however, that the on-off switch is identical to other switches, it is located next to the other switches, and the operator must turn off one of the adjacent switches. In this case, the probability of the human error "Power switch in OFF position" is relatively high, because it would be easy for the operator to turn off the power switch when he or she intended to operate the adjacent switch. A probability of 1×10^{-1} is assigned to that event, considering the identical design of the switch and its location.

In another normal event situation, the failure analysis team may be concerned about pressure spikes exceeding the maximum permissible pressure level. To quantify the probability of this occurring, the failure analysis team could instrument the system to determine how often the pressure spikes exceed the maximum level. If the instrumentation shows this occurs 2% of the time, the probability of this normal event is 2×10^{-2}. The failure analysis team may simply estimate this probability without the instrumentation. The approach taken (testing versus estimating) depends on the criticality of the pressure spike occurring. This is something the failure analysis team must decide when assigning a probability.

Basic Failure Events. The probability of a basic failure event occurring is determined using the equation presented earlier:

$$P_F = 1 - e^{-\lambda t}$$

The operating time to be used in this equation can either be specified by the customer, or the failure analysis team can estimate it.

The failure analysis team can define the failure rate based on published failure-rate sources, by contacting the component supplier, or by making an estimate based on similar components. Most electronic component suppliers provide either the failure rate or the MTBF (one is simply the inverse of the other). MIL-HDBK-217F, *Reliability Prediction of Electronic Equipment* (Ref 2), is the accepted standard for electronic component failure rates. This document is based on decades of failure-rate data collection by the U.S. Department of Defense. Failure rates for mechanical components are harder to find, but mechanical component failure-rate sources do exist.

Undeveloped Event Probabilities. Recall that an undeveloped event represents an event the failure analysis team does not wish to develop further. This usually occurs for one of two reasons:

- The undeveloped event is considered highly unlikely and therefore requires no further development.
- The undeveloped event represents failure of a subassembly for which the failure analysis team does not have any internal design information.

In the first situation, the failure analysis team can assign a very low probability of occurrence to the undeveloped event (perhaps 1×10^{-7} or 1×10^{-8}). The danger in doing this is that the probability of occurrence may be higher.

In the second situation, the failure analysis team can sometimes determine the probability of occurrence by contacting the subassembly supplier. In some cases, the supplier may be able to provide the subassembly MTBF, which can then be converted to a failure rate and used in the equation defined previously.

INHIBIT Probabilities. INHIBIT condition probabilities are usually estimated based on engineering analysis. Consider the wheel well example in Chapter 6, in which the tire has a blowout, and tire fragments rupture a hydraulic line. To calculate the probability of a tire fragment striking a hydraulic line, the failure analysis team can determine the total area of the wheel well and the total presented area of the hydraulic lines in the wheel well. In this situation, the probability of a single tire fragment striking a hydraulic line, given that the tire bursts, is:

$$P_{Strike} = \text{(Hydraulic line presented area)/(Total wheel well surface area)}$$

Command Event Probabilities. Chapter 5 stated that:

- A command event above an OR gate occurs if any of the events going into the OR gate occurred.
- A command event above an AND gate occurs if all of the events going into the AND gate occurred.
- A command event above an INHIBIT gate occurs if the event below the inhibit gate occurred and the inhibiting condition was satisfied.

The probability of a command event occurring above an OR gate is equal to the sum of the probabilities of each event immediately below the OR gate, minus the products of all combinations of the probabilities going into the OR gate. That sounds complicated. It is easier to understand with an example.

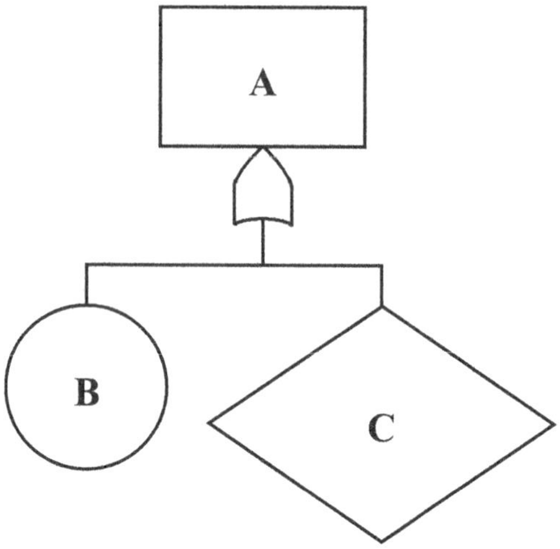

Fig. 7.1 Command event above an OR gate. The probability of occurrence of A is equal to the probability of B plus the probability of C, minus the probability of B times the probability of C.

The simplest situation is one in which only two events go into an OR gate, as shown in Fig. 7.1.

Assume the system being analyzed has an operating time of 2 hours, the component failure shown in B has a failure rate of 2.2×10^{-7}, and the subassembly failure shown in C has a failure rate of 1.6×10^{-6}. In this case, the probability of events B and C can be calculated using the failure rate and the 2 hours operating time:

$$P_B = 1 - e^{-\lambda t} = 1 - e^{-(2.2\times10^{-7})(1)} = 4.4 \times 10^{-7}$$

$$P_C = 1 - e^{-\lambda t} = 1 - e^{-(1.6\times10^{-6})(1)} = 3.2 \times 10^{-6}$$

The probability of command event A occurring is therefore:

$$P_A = P_B + P_C - P_B * P_C$$
$$= (4.4 \times 10^{-7}) + (3.2 \times 10^{-6}) - (4.4 \times 10^{-7}) * (3.2 \times 10^{-6}) = 3.64 \times 10^{-6}$$

The situation becomes more complex when considering more than two inputs into an OR gate. As stated previously, the probability of the command event occurring above an OR gate is equal to the sum of the probabilities of each event immediately below the OR gate, minus the products of all combinations of the probabilities going into the OR gate. If three events, B, C, and D, went into an OR gate below command event A, for example (Fig. 7.2), the probability of command event A occurring is:

$$P_A = P_B + P_C + P_D - (P_B * P_C + P_C * P_D + P_B * P_D + P_B * P_C * P_D)$$

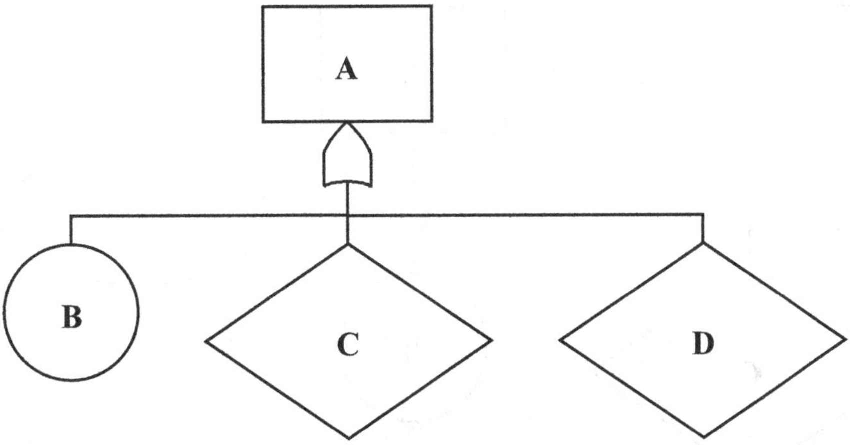

Fig. 7.2 Command event above an OR gate with three inputs. The probability of occurrence of A is given by the formula in the text.

Most of the time (particularly when describing component failure rates and their resulting calculated probabilities), the probabilities of component failures are relatively low. Therefore, the product of these probabilities becomes even smaller (for example, the product of two probabilities of 1×10^{-6} is 1×10^{-12}). As a practical matter when performing the previous calculations, the failure analysis team can ignore the product terms and simply sum the individual event probabilities.

Calculating the probability of a command event above an AND gate is relatively simple. The probability of this command event is simply the product of the probabilities of the events going into the AND gate. Suppose two events are going into an AND gate, as shown in Fig. 7.3. The probability of command event A occurring is:

$$P_A = P_B * P_C$$

If an AND gate has more than two events going into it, the probability of the command event above the AND is simply the product of all of the events going into the AND gate. There are no additional permutations or combinations of events, as when calculating the probability of an OR gate situation.

The probability of a command event occurring above an INHIBIT gate is similarly simple. In this case, the probability is equal to the probability of the event going into the INHIBIT gate times the probability of the inhibiting condition occurring. Figure 7.4 shows such a situation, where the probability of command event A occurring is:

$$P_A = P_B * P_C$$

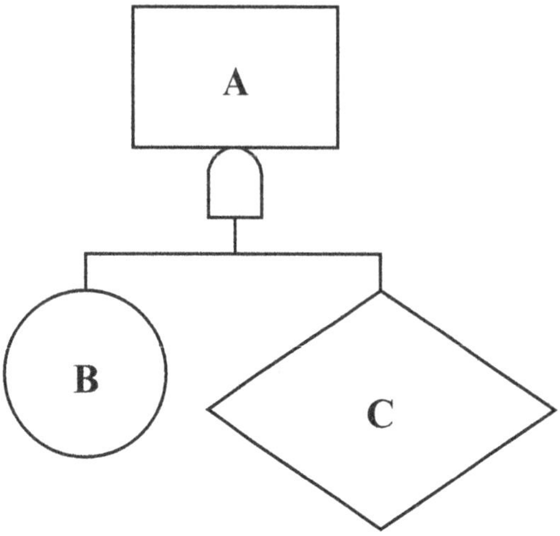

Fig. 7.3 Command event above an AND gate. The probability of occurrence of A is the product of the probabilities of B and C.

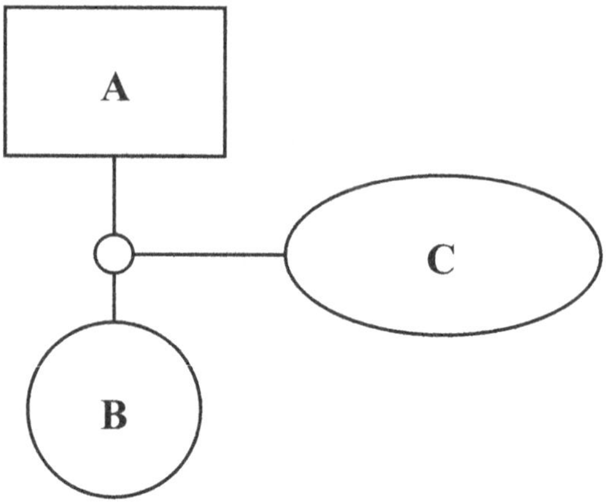

Fig. 7.4 Command event above an INHIBIT gate. Similar to the AND gate situation, the probability of occurrence of A is the product of the probabilities of B and C.

Calculating Top Undesired Event Probabilities

The top undesired event in the fault tree is the problem that the failure analysis is focused on solving. Calculating the probability of the top undesired event is accomplished using the following process:

- The failure analysis team starts at the bottom of the fault-tree analysis, quantifying the probability of each basic failure event, undeveloped event, normal event, and human error.
- The failure analysis team moves up one level from the basic, undeveloped, normal, and human error events to quantify the probability of each command event. These probabilities are calculated based on whether the command event is above an AND, OR, or INHIBIT gate.
- The failure analysis team continues moving up through the fault-tree analysis, similarly quantifying the probabilities of command events higher up in the fault-tree analysis, until the probability of the top undesired event has been quantified.

In addition to quantifying the probability of the top undesired event, the failure analysis team has also quantified the probabilities of each fault-tree branch. By so doing, the failure analysis team has identified the most likely combinations of failures that can induce the top undesired event.

In most cases, one need not be concerned about obtaining a precise probability of occurrence for each failure cause or an exact failure rate for the basic failure events. There is a phenomenon called the "big number" theory, which means that some of the estimates will be high, some will be low, but in the aggregate, the overall estimate will average out the high and low values.

Fault-Tree Quantification and Systems Failure Analysis

Quantifying a fault-tree analysis is a useful tool for assessing the most likely causes of a system failure. In most cases in which fault trees are quantified, the quantification is accomplished to use the fault tree as a predictive tool. During aircraft development efforts, for example, clients require that probability of aircraft loss rate not exceed a specified value. The ability of the fault tree to quantify the probability of such aspects allows a client or an aircraft developer to predict if the overall design meets its requirements. If a fault-tree analysis shows that the probability of loss exceeds a specified limit, the engineering team can review the fault-tree branches to identify areas contributing significantly to the unacceptably high loss rate and redesign the system accordingly.

For example, during the F-16 development program, the U.S. Air Force required that the F-16 not experience more than one aircraft loss every 100,000 flight hours. During development, a fault-tree analysis showed that this overall system failure rate would be exceeded. General Dynamics redesigned portions of the system until the aircraft design resulted in a predicted loss rate below the maximum allowed. As a point of interest, the actual aircraft loss has been substantially below 1×10^{-5} losses per flight hour, demonstrating the validity of this approach.

In a systems failure analysis context, fault-tree quantification can be useful for identifying the most likely failure causes. In the systems failure analysis context, the failure has already occurred. The objective in preparing the fault-tree analysis in this situation is to identify all potential failure causes, initially without regard to which are likely and which are unlikely. That assessment occurs later, and the techniques for making these determinations are the topics of the next several chapters. This point is particularly relevant in the context of a recurring failure (one of the hidden factory failures mentioned in Chapter 2). In these situations, the failure causes are probably obscure and not recognized as likely causes. If the causes were obvious, they likely would have been eliminated long ago.

After completing the fault-tree analysis and converging on the actual failure causes, the failure analysis team is left with many other potential causes. It may not be practical to implement corrective actions to prevent all of these from occurring, so some means are needed of quantifying which ones are likely to occur and which ones are not. By quantifying the fault-tree analysis, the failure analysis team can implement corrective actions where it makes sense to do so and intelligently ignore areas in which the probability of a failure is sufficiently low.

Summary

Fault trees can be quantified to identify the probability of occurrence of the top undesired event. Basic event failure rates can usually be found in published failure-rate sources, or they can be estimated. The failure analysis team can convert these failure rates into probabilities of occurrence. Similarly, undeveloped event probabilities can be based on supplier MTBFs or engineering estimates. Normal events and inhibiting conditions are usually based on engineering analysis. Human error probabilities can be based on published sources or engineering judgment. Human error probabilities vary widely based on design features, stress levels, and operational complexity. Command event probabilities are based on the gate type beneath the command event and the probabilities of the events below the command event. The OR gate probabilities are complex, but these can be simplified as the sum of the event probabilities going into the OR gate if the individual probabilities are relatively low. The AND and INHIBIT gate probabilities are the products of the events and conditions going into the gate.

Fault-tree analysis quantification is useful for determining both the probability of the top undesired event and the probabilities of individual fault-tree branches. In a systems failure analysis context, it is important not to rule out less probable failure causes. All potential failure causes should be objectively evaluated, even if their predicted probability of occurrence is low. Fault-tree quantification can be a useful tool for deciding where to implement corrective actions for hypothesized failure causes that have not yet occurred but are predicted to be likely to do so.

Example for Group Discussion. In a systems failure analysis context, the systems failure has already occurred. Based on this, is the probability of the top undesired event 100%?

Under what circumstances may an organization want to prepare a fault-tree analysis for a hypothesized systems failure (even though no failure has occurred)?

REFERENCES

1. B. Kirwan, *A Guide to Practical Human Reliability Assessment,* Taylor & Francis, Ltd., London, 1994
2. *Reliability Prediction of Electronic Equipment,* MIL-HDBK-217F, U.S. Department of Defense

SELECTED REFERENCES

- I. Bazovsky, *Reliability Theory and Practice,* Prentice-Hall, Englewood Cliffs, NJ, 1961
- *Fault Tree Construction Guide,* Armament Development Test Center, U.S. Air Force, 1974
- W.F. Larsen, *Fault Tree Analysis,* Ammunition Reliability Division, Picatinny Arsenal, U.S. Army, 1968

CHAPTER **8**

Failure Mode Assessment and Assignment

APPROACHES FOR IDENTIFYING potential failure causes have been described in this book, thus far. These have included brainstorming, mind mapping, Ishikawa diagrams, the "five whys" technique, flow charting, and fault-tree analysis (which, as previously noted, is preferred). After using one or more of these techniques to identify potential failure causes, the failure analysis team may have hundreds or even thousands of potential causes to assess. The failure analysis team now needs a tool to help manage the evaluation of each hypothesized failure cause.

Failure mode assessment and assignment (FMA&A) is a tool designed to help organize the evaluation of hypothesized failure modes. The FMA&A is a four-column table that shows:

- Event number of each hypothesized failure cause
- Description of each hypothesized failure cause
- Likelihood assessment of each hypothesized failure cause
- Actions necessary to evaluate each hypothesized failure cause and the status of each evaluation

To assist in describing the preparation of an FMA&A, the fault-tree analysis described earlier in Chapter 5 for the top undesired event, "Light bulb does not illuminate," is used once again. For easy reference, the light bulb fault-tree analysis is included in this chapter as Fig. 8.1. Table 8.1 shows the FMA&A associated with the light bulb fault-tree analysis.

FMA&A Preparation

After the failure analysis team has prepared a fault-tree analysis (or used one of the other methods to hypothesize potential failure causes), the team should prepare an FMA&A to assign actions. These actions focus on identifying

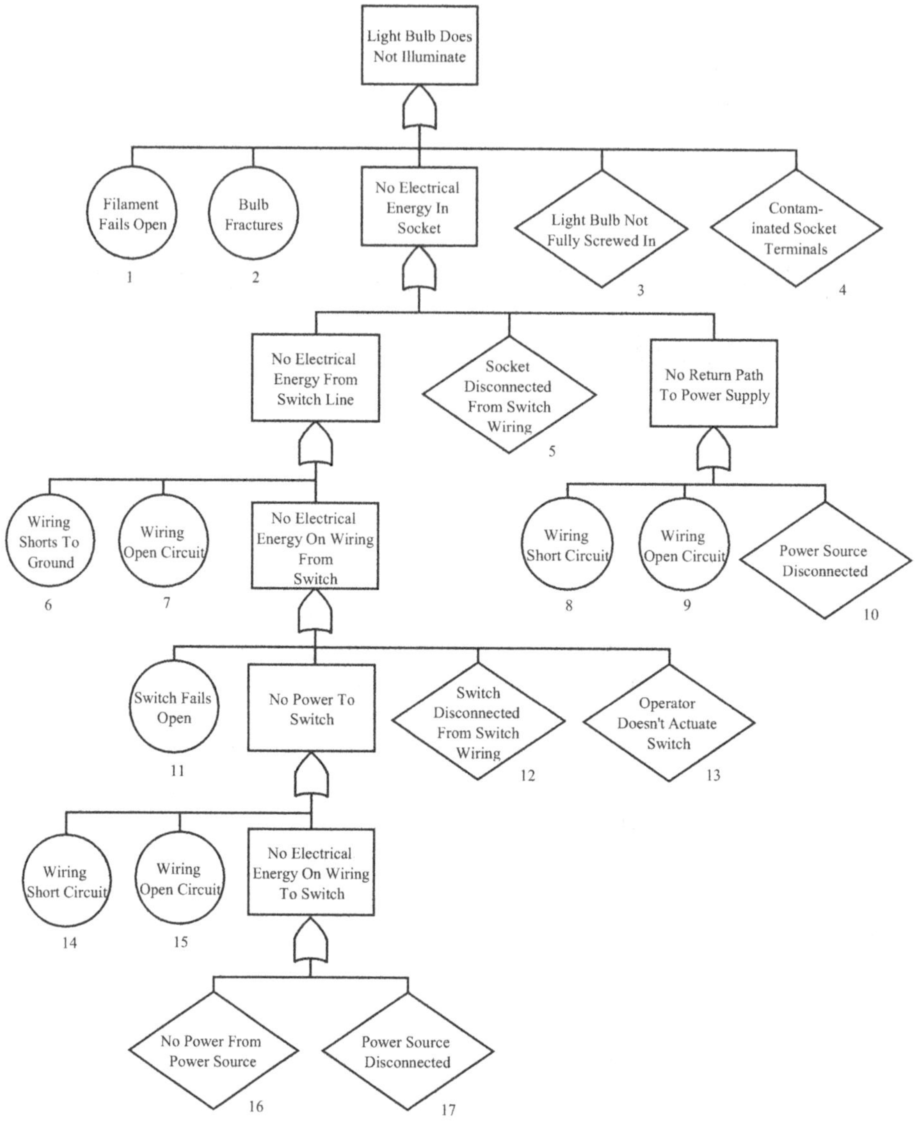

Fig. 8.1 Indicator light fault-tree analysis develops potential causes for the light bulb failing to illuminate. The development of this fault-tree analysis is described in Chapter 5.

which of the hypothesized failure causes induced the failure. An efficient way to do this is to prepare the FMA&A using either a word-processing or spreadsheet program, with the team members participating in a determination of the actions necessary to assess each hypothesized failure cause. Each column of the FMA&A should be completed according to the following outline.

Event, Sheet Column. The first column in the FMA&A is the event- and sheet-numbering column. This column ties each FMA&A row to an event in the fault tree. The failure analysis team should only assign event numbers to basic failures, undeveloped events, human errors, normal events,

Table 8.1 Indicator light failure mode assessment and assignment (FMA&A) matrix

The FMA&A shows the actions necessary to evaluate each potential failure cause identified by the fault-tree analysis. The gray-shaded rows indicate that the actions defined in that row have been completed.

Event, sheet	Hypothesized failure cause	Assessment	Assignment
1, 1	Filament fails open	Unlikely	Examine bulb for open filament. Rodriguez; March 14. Action completed. Bulb examined and tested; filament intact
2, 1	Bulb fractures	Unlikely	Examine bulb for open filament. Rodriguez; March 14. Action completed. Bulb examined and tested; bulb intact
3, 1	Light bulb not fully screwed in	Unlikely	Inspect bulb in socket to determine if properly installed. Smith; March 14. Action completed. Smith checked light bulb and found it was fully seated.
4, 1	Contaminated socket terminals	Unknown	Examine socket for contaminants. Perform Fourier transform infrared analysis on any contaminants observed in socket. Rodriguez; March 16
5, 1	Socket disconnected from wiring	Unknown	Examine wiring and perform continuity test. Ashoggi; March 16
6, 1	Wiring shorts to ground	Unknown	Examine wiring and perform continuity test. Ashoggi; March 16
7, 1	Wiring open circuit	Unknown	Examine wiring and perform continuity test. Ashoggi; March 16
8, 1	Wiring shorts to ground	Unknown	Examine wiring and perform continuity test. Ashoggi; March 16
9, 1	Wiring open circuit	Unknown	Examine wiring and perform continuity test. Ashoggi; March 16
10, 1	Power source disconnected	Unknown	Determine if connections to power source are good. Smith; March 17
11, 1	Switch fails open	Unknown	Remove switch and test in lab, including repeated cycling to determine if intermittent condition exists. Smith; March 18
12, 1	Switch disconnected from wiring	Unknown	Determine if connections to switch are good. Smith; March 17
13, 1	Operator does not actuate switch	Unknown	Interview all operators and observe operator function system. Nelson; March 18. Review operating instructions for clarity and directions related to switch actuation. Nelson; March 18
14, 1	Wiring shorts to ground	Unknown	Examine wiring and perform continuity test. Ashoggi; March 16
15, 1	Wiring open circuit	Unknown	Examine wiring and perform continuity test. Ashoggi; March 16
16, 1	No power from power source	Unlikely	Check power supply with multimeter. Ashoggi; March 14, 1993. Action complete. Power source tested and found to be operational.

and inhibiting conditions. The team should not assign event numbers to command events, because these are all induced by one or more of the other previously listed events. If the fault-tree analysis is comprised of more than one sheet (and most are), it is a good idea to also include the fault-tree sheet number in this column. (The fault-tree example included here has only one sheet for the sake of brevity, but the sheet number is included in the FMA&A to demonstrate the concept.)

If methods other than the fault-tree analysis have been used to hypothesize potential failure modes, the failure analysis team can simply number the potential failure causes from "1" to the total number of causes.

Hypothesized Failure-Cause Column. The second column in the FMA&A shows an event description for each hypothesized failure cause.

Each row of the FMA&A is associated with an event (other than a command event) depicted in the fault tree. The entry in this column repeats the event description for each basic, undeveloped, human error, normal event, and inhibiting condition in the fault tree. It may seem redundant to repeat the event descriptions, but after the fault-tree analysis is complete, the failure analysis team works primarily with the FMA&A. Having the event description readily available in the FMA&A saves time and minimizes confusion.

If the failure analysis team has not prepared a fault-tree analysis and is instead relying on one of the other potential failure-cause development methods described in Chapter 4, the hypothesized failure causes are similarly listed in this column.

Assessment Column. The assessment column shows the failure analysis team's assessment of each hypothesized cause. When the failure analysis team prepares the initial FMA&A draft, the team does not yet know which of the hypothesized failure causes induced the failure. Based on this, each hypothesized failure-cause assessment is initially listed as "unknown." As the failure analysis team completes actions evaluating the likelihood of each failure, these ratings may change to "unlikely," "likely," or "confirmed."

The likelihood assessments are relatively subjective. "Unlikely" means that the failure analysis team objectively evaluated the hypothesized failure cause and concluded that it was unlikely to have induced the failure. "Likely" means the team concluded that a hypothesized failure cause was likely to have induced the failure, but the evidence is not conclusive. "Confirmed" means the failure analysis team evaluated the hypothesized failure cause and found objective evidence that the cause was present and induced the failure.

The failure analysis team should not allow the qualitative assessment of this column to be exclusively determined by the fault-tree analysis event probability, as discussed in Chapter 7. If event probabilities are developed during the fault-tree analysis, they can be used to aid in assessment of each hypothesized failure mode, but the failure analysis team should not use an event probability by itself to assess which of the hypothesized failure modes induced the failure. For this, the team must identify specific actions to evaluate the likelihood of occurrence of each hypothesized failure cause. These actions are listed in the fourth FMA&A column, as shown in Table 8.1.

Assignment Column. The fourth FMA&A column defines actions necessary to evaluate each of the hypothesized failure causes. The failure analysis team prepares the initial draft of the FMA&A matrix by completing the first three columns described previously (using inputs from the fault-tree analysis or any of the other methods described earlier). This activity is mechanical in nature and proceeds relatively quickly. If the fault-tree analysis is in a computerized format, the team can further accelerate the process by having the fault-tree analysis and table files open and simply copying and pasting the event descriptions from one to the other.

After the first three columns of the FMA&A are prepared, the failure analysis team reviews the FMA&A on a row-by-row basis. During this review, the failure analysis team identifies actions necessary to objectively evaluate the likelihood that each hypothesized failure cause induced the system failure being analyzed, as shown in Table 8.1.

The following guidelines are effective when assigning actions to evaluate each hypothesized failure cause:

- The failure analysis team discusses each hypothesized failure cause fully and focuses on defining actions necessary to rule the hypothesized cause in or out. If all of the failure analysis team members are included in this discussion, the methods proposed for making this assessment are more meaningful. Some of the failure analysis team members may wish to subjectively eliminate selected hypothesized failure causes by changing their assessment from "unknown" to "unlikely." In some cases, this may make sense, but the team must review such decisions carefully. Recall that in the case of recurring failures, the causes are not going to be obvious. (If they were, they would already have been corrected.)
- When assigning actions, the failure analysis team assigns only one name to each action. Experience indicates that if multiple names are assigned to the same action, the action is less likely to be completed.
- The failure analysis team negotiates a completion date with the person assigned to each action and includes this information in the assignment column.

After the initial FMA&A draft has been prepared, the failure analysis team can focus on follow-on activities to evaluate each hypothesized failure cause.

Follow-On Activities

At this point, it is appropriate to emphasize a point developed earlier in this book: The failure analysis team should not touch any hardware from the failed system until the team has defined an organized, systematic, and objective manner in which to proceed. The FMA&A provides this organized, systematic, and objective plan.

After the initial FMA&A draft has been prepared, the failure analysis team members know what they need to do to rule in or out hypothesized failure causes. After the initial meeting, the team members execute these FMA&A-required actions and report their findings in subsequent meetings. As the hypothesized failure mode evaluations are completed, the findings are included in the FMA&A assignment column. As indicated by the team's findings, the assessment is changed to show if the assessment of the hypothesized cause changes. When these actions are completed, the row in which they appear is shaded gray to quickly indicate which actions are completed and which hypothesized failure causes are still under evaluation.

Failure analysis team meeting frequency depends on the severity and consequences of the failure being analyzed. If the team is examining significant failures (those with important cost, safety, or customer impact), it is recommended that the team meet at least daily. Failure analysis team meetings for other types of failures may occur less frequently. In all cases, it is recommended that the meetings occur at least weekly. The FMA&A is used to guide the meeting. It is recommended that the FMA&A be updated and distributed to all team members and other interested parties after each failure analysis team meeting.

Evaluating Potential Failure Causes

The analysis techniques covered thus far have been focused on steps 1 and 2 of the four-step problem-solving process (Fig. 1.2). Potential failure causes have been developed, and this chapter presented an organized approach for managing the evaluation of hypothesized potential causes. It is now necessary to address how to perform these evaluations. These topics, covered in the next several chapters, include:

- Pedigree analysis
- Change analysis
- Analytical equipment
- Mechanical and electronic component failures
- Leaks
- Contamination
- Design analysis
- Statistical considerations
- Design of experiments

After discussing the aforementioned topics, the last two steps in the four-step problem-solving process are addressed (i.e., identifying potential solutions and selecting the best solution). In the failure analysis context, this is called corrective action.

Summary

The failure analysis team identifies all potential failure causes when analyzing a system failure and uses these to prepare an FMA&A matrix.

When the potential failure causes have been identified and listed in the FMA&A, it provides a tool for managing follow-on actions. These actions objectively evaluate each hypothesized failure cause to determine if it caused the failure or could cause similar failures in the future. The failure analysis team uses the FMA&A to manage the failure analysis and updates and distributes the FMA&A after each failure analysis team meeting.

Example for Group Discussion. As mentioned earlier, the failure analysis team should resist the temptation to disassemble and examine failed products immediately after the failure is discovered. The failure analysis team should develop a fault-tree analysis and an FMA&A showing all potential failure causes and the actions required for each cause before disassembling and examining the failed system, which may contain valuable clues to the causes of the failure. Tearing the failed system apart immediately after a failure without knowing what to search for (i.e., what may have caused the failure) risks destroying valuable evidence. However, many organizations disassemble failed systems and look for the failure cause prior to identifying what may have caused the failure.

Which approach does your organization follow? Do you think your failure analysis approach needs to change? If so, what can you do to initiate a change?

SELECTED REFERENCES

- S. Berk and J. Berk, *Quality Management for the Technology Sector,* Newnes, 2000
- L. Stewart and J.H. Berk, Ordnance System Failure Analysis, *Proceedings of the Annual Reliability and Maintainability Symposium,* 1987

CHAPTER 9

Pedigree Analysis

PRODUCTS HAVE PEDIGREES. A product pedigree describes its design and how it was built and shows that it was built in accordance with the drawings and other documentation defining the product configuration. A product pedigree is comprised of engineering drawings, specifications, process descriptions, test results, inspection data, and other documentation. Evaluating the pedigree of a failed product can help to rule in or rule out hypothesized failure causes.

The documents making up a product pedigree are records of how it was built, the materials used, and conformance to drawing and material requirements. As such, the failure analysis team can use these documents to assess if any hypothesized failure causes were present in the failed system. The technique is to consider each hypothesized failure cause listed in the failure mode assessment and assignment (FMA&A) and ask if the suspect condition may be revealed by an analysis of the product pedigree. This is valuable from two perspectives:

- If the pedigree addresses the suspect area, it can be examined to assess if the pedigree reveals the presence of the suspect condition.
- If the pedigree does not address the suspect area, the failure analysis team may consider recommending that it do so on future builds as a corrective action to prevent future failures.

The failure analysis team can examine the following areas to assess the pedigree of the failed system:

- *Purchase orders:* The failure analysis team should review the client's purchase order for the product as well as any purchase orders for supplier items used in the system. This determines if the product is what the client ordered and if all purchased items used in the system are what the system design required. In some cases, ordering or using incorrect products for the intended application can cause failures. If such a situation exists, the failure analysis team can discover it by reviewing all relevant purchase orders. To the extent that it is possible to do so, the failure analysis team should attempt to review their suppliers' purchase orders in areas where failure causes can affect system performance. Attempting to review such information is inherently difficult. (Suppliers are reluctant to share subtier purchase orders with their customers, because this will reveal cost information that the supplier probably does not want to share.) Including a purchasing representative on the failure analysis team can help in this area.

- *Nonconformance documentation:* Most organizations document all of their nonconformances (i.e., areas not meeting drawing requirements) on nonconforming material reports. This is especially true in highly regulated industries, such as aerospace, defense, biomedical, cosmetics, and food manufacturers. These records are retained in the organization's quality-assurance files and database. The failure analysis team should review these records for nonconformances, and the team should do so both within the organization and at suppliers providing items that could have caused the failure. (This is another reason for including purchasing personnel on the failure analysis team.) Readers may wonder why the failure analysis team must review such records. There are several reasons. One is that rejected items sometimes find their way back into production without the rejectable condition having been addressed. Even if the part from the failed system was not rejected, the failure analysis team may find rejections on other parts (suggesting the item in the failed system may be an inspection escape). The failure analysis team may find sloppiness or errors in the inspection documentation, again suggesting an inspection escape.
- *Inspection records:* In a manner similar to that described previously, most companies retain their inspection records to show that the products and materials they used met all requirements. The failure analysis team should review the inspection records in areas related to the hypothesized failure causes. The data may show that requirements were not met, but the product continued through production as if it conformed to the requirements. The inspection data may show that the requirements were just barely met. (In other words, the product dimensions or other characteristics were just inside the allowed tolerance bands.) This suggests that other product may have exceeded the allowed tolerance bands due to normal statistical variation, especially if the supplier inspected on a sampling (rather than 100%) basis. Finally, the inspection records may show attributes data (e.g., "acceptable" or "pass" notations) instead of recording variables (or numerical) data. This also suggests that the product could be outside allowed dimensional tolerances due to inspection error, calibration error (discussed subsequently), or normal statistical variability if the inspection was performed on a sampling basis.
- *Test data:* In addition to dimensional inspections, manufacturers and suppliers often test materials, components, and subassemblies. The

failure analysis team should review test data for the presence of failure causes hypothesized by the FMA&A, considering the same factors outlined previously.

- *Calibration data:* Equipment used for acceptance of product and materials should be calibrated to ensure that it is accurate. The failure analysis team should review calibration records for all test and inspection equipment wherever an out-of-calibration condition may induce a hypothesized failure cause. Sometimes, such reviews reveal that the inspection or test equipment is overdue for calibration. This does not necessarily mean that the item is out of calibration, but it is an area that suggests a problem. Even if the calibration is not overdue, if it is in a critical area, the failure analysis team may consider evaluating the accuracy of the inspection equipment. This is particularly true if the item has a lengthy period between calibrations, or if it is likely that the inspection tool is out of calibration due to being dropped or otherwise abused.
- *Drawings and specifications:* Products are built to engineering drawings and, sometimes, to specifications. The failure analysis team should check that the correct drawings and specifications were used. This should be accomplished both internally and at suppliers.
- *Drawing changes:* Records of changes to drawings and specifications are usually kept in a document control or engineering department of an organization. The failure analysis team should check to see if required changes have been implemented or if changes not required have been erroneously implemented. This should be accomplished both internally and at suppliers.
- *Work instructions:* As mentioned previously, product designs are typically documented in drawings and specifications. Changes to these documents are usually under some form of change control. The same is not always true for work instructions. (In fact, in less highly regulated industries, work instructions may not exist at all.) To the extent that it is possible to do so, the failure analysis team should check work instructions (both internally and at any affected suppliers) to determine if the product was built using the correct process.
- *Certificates of conformance:* In many instances, manufacturers do not have raw material, component, or dimensional data on items provided by suppliers. Manufacturers instead rely on a certificate of conformance from the supplier. A certificate of conformance states that all requirements have been met. This typically includes material, process, dimensional, and, sometimes, performance requirements. Relying on certificates of conformance is a common industry practice based on an assumption that the certificate is truthful and accurate. Economically, it makes sense to do this, because it would add unnecessary cost and a duplication of effort to products if both the supplier and the buyer inspected the product. Certificates of conformance for

supplier-purchased items are a key part of a product pedigree. The failure analysis team should review the product certificates of conformance in potential failure-cause areas to assess if the certificates suggest that any of these causes were present.

The preceding areas typically define the product pedigree. Sometimes, a question arises regarding why it is necessary to check this information, because it intuitively seems that such information would most likely indicate the system meets all of its requirements. Surprisingly, this is often not true. From past experience, pedigree reviews frequently reveal that the product does not meet all of its requirements. In some cases, the nonconforming areas are directly related to the failure being analyzed.

It is not necessary to check every aspect of a product pedigree. It is only necessary to check those areas in which hypothesized failure causes listed by the FMA&A can induce the system failure.

When reviewing the product pedigree, the failure analysis team should recognize that errors, omissions, or fraud may make the pedigree suspect. Just because a supplier provides a certificate of conformance does not guarantee that a product meets its requirements, as the following case illustrates.

A Tragedy in Hawaii

Several years ago, an elderly couple celebrating their 50th wedding anniversary took a private aerial tour of the Hawaiian Islands. The couple hired a company offering private flights in a small, twin-engine aircraft. Shortly after takeoff, one of the two engines on the aircraft caught fire. The pilot lost control of the airplane; it crashed into the ocean, and the woman died. The team investigating this accident found that oil leaking onto an engine caused the fire. The oil leaked because a rubber oil filter gasket had melted.

The husband sued the tour company, the airplane manufacturer, the engine manufacturer, the oil filter manufacturer, the gasket distributor, and the gasket manufacturer. As the litigation and accident investigation progressed, the following facts emerged:

- The failure cause was an oil filter gasket fabricated from the wrong material. The gasket melted, which allowed oil to run onto the engine exhaust manifold. The oil ignited, and the resulting fire caused the crash.
- Just prior to the accident, maintenance technicians had changed the oil and oil filters in the aircraft engines. During the oil change, they used an oil filter with the incorrect gasket material.
- The oil filter supplier bought components from various distributors and assembled them into oil filters. The oil filter company's drawings specified all components and materials, including the correct rubber for this application.

- The gasket distributor ordered the gasket from the gasket manufacturer using their drawing, which specified the material.
- The gasket manufacturer cut the gasket to the required dimensions using on-hand material and shipped it to the gasket distributor. The gasket manufacturer identified on its packing slip that a different material had been used.
- The gasket distributor provided the gasket to the oil filter company without specifying the gasket material.
- The oil filter company noticed that the gasket distributor did not include a certificate of conformance for the gasket. The oil filter company asked the gasket distributor for a certificate of conformance.
- The gasket distributor prepared a certificate of conformance stating that the gasket used the correct material, even though the gasket manufacturer's packing slip showed that a different material had been used. The gasket distributor sent this erroneous certificate of conformance to the oil filter company.
- The oil filter company, satisfied that the gasket met all requirements (based on the erroneous certificate of conformance), supplied the certificate of conformance and the oil filter to the tour company.
- The tour company used the oil filter, and the accident resulted.

In this case, the product pedigree and, in particular, the certificate of conformance for the oil filter gasket were inconsistent with the failed item observations. Noting this, the team took a hard look at the gasket manufacturer, the gasket distributor, and the oil filter supplier. The failure analysis team uncovered the chain of events outlined earlier.

The author has frequently encountered anomalous certificates of conformance (as well as other documentation anomalies) during the course of a pedigree review. To the author's knowledge, very few instances of outright fraud (as may have occurred in the previous example) have been encountered. It is not unusual to uncover erroneous documentation. Most often, the cause is human error. In some cases, the documentation clearly indicates nonconforming material, incorrect product types, and so on, but the item shipped anyway. If the failure analysis team suspects an area indicated as acceptable by the product pedigree, the team should seek independent verification that the suspect area meets requirements. This can be accomplished through additional inspection, testing, observation, or any of several other techniques discussed in the next several chapters.

Summary

Products are defined by the designs, specifications, work instructions, purchase orders, test data, inspection data, and other documents used in their creation. This documentation trail is referred to as the product pedigree.

When evaluating hypothesized failure causes, a review of the product pedigree can help to rule in or out hypothesized failure causes.

Product pedigrees may not show any anomalous conditions, but if the failure analysis team suspects the pedigree, it should confirm conformance through independent means. In some cases, the pedigree shows an anomalous condition, but the product shipped anyway. These situations can occur due to human error. In many cases, having someone from the organization's purchasing department on the failure analysis team helps in obtaining pedigree information from suppliers.

Example for Group Discussion. If a product pedigree shows that it is nonconforming, it should not be shipped until the nonconforming condition is addressed. In some cases, however, the product ships without correction of the nonconforming condition. How can this happen? What can be done to prevent it?

Does your organization have any other areas (in addition to those mentioned previously) that can be reviewed to ascertain the product pedigree? Do your suppliers have any such areas?

Sometimes, suppliers are reluctant to share pedigree data. How can this situation be prevented?

SELECTED REFERENCES

- S. Berk and J. Berk, *Quality Management for the Technology Sector,* Newnes, 2000
- *Quality Management Systems: Requirements,* ANSI/ISO/ASQ Q9001-2000, American Society for Quality, 2000
- "Quality Program Requirements," MIL-Q-9858, U.S. Department of Defense, 1985

CHAPTER **10**

Change Analysis

WHEN A FAILURE OCCURS, it is almost intuitive to assume that because the system had been operating without failure in the past, something must have changed. In some cases, this is true; in other cases, it is not. This chapter targets areas that determine if a change occurred and if the change induced the failure. This is referred to as change or "what's different" analysis.

Before delving into areas in which a failure-inducing change may have occurred, it is important to recognize the following possibilities:

- Perhaps nothing changed. The failure may be due to something unrelated to any change.
- Perhaps the failure has been occurring all along but has not been observed or noted in the past. Sometimes, this occurs when a new inspector is assigned to an area. Such a situation is disconcerting, because it suggests that failures have occurred in the past and were not detected.
- Perhaps the failure is occurring as the result of normal statistical variation. This may occur, for example, in the case of a mechanical failure if the upper end of the load distribution overlaps the lower end of the strength distribution.
- Perhaps a change occurred, but the change is unrelated to the failure. When searching for changes that could have induced the failure, the failure analysis team should focus on areas identified in the failure mode assessment and assignment (FMA&A). Changes in designs, processes, tooling, operators, and environment occur almost constantly, so it is likely that the failure analysis team will find changes that occurred at approximately the same time as the failure. However, just because a change occurred around the time of the failure does not mean that it induced the failure.

With the aforementioned cautions in mind, the failure analysis team should consider the following areas when searching for changes.

Sources of Change

Design changes are usually noted in drawing or specification revisions, so if a design change occurs, finding it is usually straightforward. The failure analysis team should review all relevant design documentation to determine if any changes related to hypothesized failure causes were present in the failed system. Usually, such information is available from the company's engineering or document control functions.

In some cases, design changes are less readily apparent. Sometimes, organizations find that a design change is necessary, and the change is implemented immediately in the factory or at the supplier with a "redlined" engineering drawing or specification. Redlining refers to an engineer using a red pencil to note design changes on the drawings in the shop. In effect, the engineer directs the shop to proceed at risk with a design change, with the intent being that the drawings will be formally revised in the near future. These days, the changes may be done with a red pencil on a hard copy of the drawing, or the redlined changes may be noted electronically. In either case, the concern is that a change occurs but is not yet reflected in formally revised drawings. For this reason, in addition to reviewing all design information, the failure analysis team should determine if any redlined drawing changes were incorporated in the product that failed.

Another area in which subtle design changes may occur (but not be shown in related drawings or specifications) is on rejected material dispositioned as "use as is" or "repair." When the quality-assurance organization rejects a component or subassembly, something must be done with the rejected material. Two possible dispositions are "use as is" or "repair." A "use as is" disposition means that someone with authority and knowledge evaluated the nonconformance and determined that it is acceptable to use the material in its nonconforming state. A "repair" disposition means that modifications were made to the nonconforming item to ameliorate (but not eliminate) the nonconforming condition.* In the case of either "use as is" or "repair" dispositions, the nonconforming items are deemed acceptable for use, but they do not conform to the engineering drawings. In effect, they are design changes not shown in the engineering drawings. For this reason, the failure analysis team should review all quality-assurance documentation associated with the failed product to determine if any "use as is" or "repair" dispositions are changes that could have induced the failure.

In most cases, a large percentage of an organization's product is made up of purchased items. Such items are frequently procured to specifications or top-level requirements, without requiring the suppliers to provide detailed design information about the inner workings of the purchased item. The types of changes described earlier (formal design changes, redlined design

---

*For example, a metal component with a crack could be dispositioned with a weld repair. The weld eliminates the unacceptable crack and makes the item functional, but the item does not conform to the engineering drawing.

changes, or "use as is" or "repair" dispositions) may be present in these items. The failure analysis team's purchasing representative should work with suppliers to determine if any changes that could induce the failure were present in the purchased items of the system.

Process Changes. Unlike design changes, uncovering process changes can be a little more elusive, for several reasons:

- Many companies do not have work instructions. If the process is not documented, changes to the process are not documented.
- In most companies that have work instructions, the work instructions are not precise. In many cases, this is desirable, because it allows for some process flexibility.* When work instructions allow for operator flexibility, minor process changes occur with no documentation.
- In most companies that have work instructions, changes to the work instructions are usually not as rigorously controlled or documented as changes to the engineering drawings and specifications.
- In companies that have work instructions, changes to tooling, production and inspection equipment, or settings on these items may not be documented.
- People do not always follow work instructions. If the work instructions are not followed, changes to the process are not documented.

The failure analysis team should evaluate the process to determine if any changes occurred that relate to hypothesized failure modes. It is almost a certainty that changes have occurred; the failure analysis team must determine if any of these induced the failure. New machinery, fixturing, setup methods, machine speeds, machinery location, mold temperatures, operator reassignments, and other factors should all be reviewed to identify changes.

As a starting point, the failure analysis team should review available process documentation to determine if any changes can be correlated to the failure being analyzed. However, for the reasons outlined earlier, it is likely that if such changes occurred, they were not documented. It is usually necessary to visit operators and suppliers to identify process changes. Caution is required here. If the failure analysis team is perceived as accusatory, elitist, or threatening, it is not likely that the operators or suppliers will be forthcoming. Failures are a touchy subject, and few people want to be perceived as having played a role in inducing a failure. A low-key, participatory approach works best. Visiting operators in their work centers (rather than pulling them into meetings), visiting on a one-on-one basis (rather than having everyone on the failure analysis team show up in the work center), and asking open-ended questions (rather than questions that can be answered with a "yes" or a "no") are approaches that work well.

---

*This is desirable because it keeps the size and length of the work instructions reasonable, and it reduces work instruction preparation and maintenance cost. Another reason is that it allows the company some protection against overly zealous process auditors in highly regulated industries.

In companies that use statistical process control (SPC), the SPC charts are helpful in identifying changes. Operators using SPC are trained to note all process changes directly on the SPC chart, so reviewing these offers insight into process changes not noted elsewhere.

Test and Inspection Changes. The failure analysis team should review test and inspection activities using the techniques and examining the areas described previously. Test and inspection activities are usually more fully documented than other manufacturing activities, so the failure analysis team will probably find more documentation that can be used to determine if changes occurred.

The beginning of this chapter mentioned the situation involving a change of inspectors, with the new inspector rejecting items more frequently than the earlier inspector. The failure analysis team must determine if a rejection associated with a new inspector is valid. Sometimes, the new inspector may not perform the inspection or test correctly, and the rejection is incorrect (i.e., the part is conforming). In other cases, the new inspector may be correct, and the rejection is legitimate. In that case, the failure analysis team has another challenge (determining if the items accepted by a previous inspector were conforming).

Another possibility is that an inspector change upstream in the manufacturing process may allow unacceptable material to pass through, which induces failures downstream. The failure analysis team should also evaluate the process for this type of change.

The failure analysis team should look for recent instrument or inspection equipment calibrations in areas that could have induced the failure. It is possible that if the calibration was erroneously performed, it allowed nonconforming material to reach a point where it could induce a failure. It is also possible that the inspection equipment erroneously accepted material prior to calibration and is now rejecting material because the inspection equipment is in calibration. The failure analysis team now has the same new challenge described earlier—determining if previously accepted material was actually nonconforming.

Environmental Changes. In some cases, environmental changes can induce failures. The failure analysis team should consider this when preparing the fault-tree analysis, as outlined in the discussion on state-of-the-part failures in Chapter 6. This may occur, for example, in processes involving drying or curing of materials, particularly if the process is performed in a nonenvironmentally controlled area. The failure analysis team may consider correlating the manufacturing date of the failed system with the relative humidity and temperature on that date.

Sometimes, how material is stored influences how it performs. The failure analysis team may wish to determine if materials associated with any of the hypothesized failure causes experienced changes in storage environments. For epoxies and composite raw materials, for example,

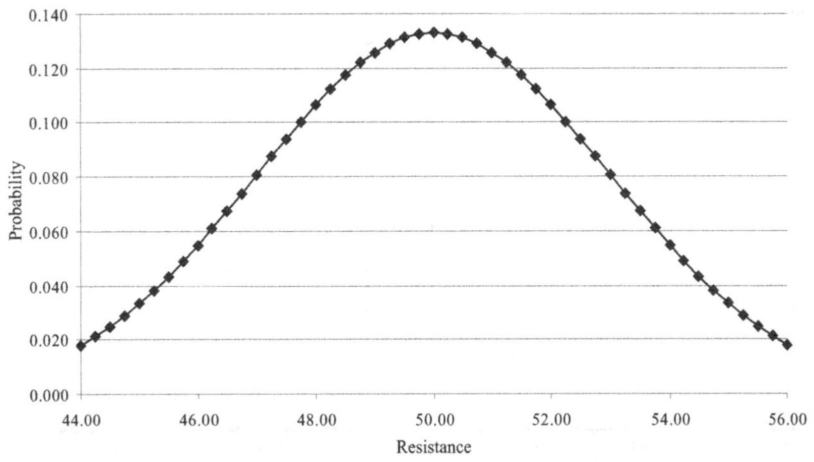

Fig. 10.1 Resistor normal distribution curve showing the distribution of values for resistors produced prior to sorting. The mean value in this example is 50 Ω, and the standard deviation is 3 Ω.

storage temperature is critical. Moving raw materials to a different storage location or incorporating changes in the current storage location may induce an environmental change that can be correlated to a hypothesized failure cause.

Lot Changes. Sometimes, supplier lot changes can be correlated to failures. If a new lot of a suspect item is introduced at the time the failure occurs, the failure analysis team should determine if something in the new lot induced the change. Because many purchased items are bought to top-level specifications (and not detailed internal design requirements), it is critical to involve the supplier. The failure analysis team's purchasing representative can help in this area.

It is also useful to have an understanding of how suppliers produce their products. Resistors, for example, are usually purchased with a nominal resistance value and a tolerance expressed as a ± percentage of the nominal value. Resistor manufacturers produce a large lot of resistors and then measure the resistance of every resistor. Some resistors are very close to the nominal value, and the distribution appears almost uniform. Some have values farther away from the nominal value, and these are used for purchase orders with looser tolerances. As the tolerance is increased, the resistors in the more loosely toleranced lots have a pronounced bimodal distribution. Figures 10.1 to 10.3 show this effect.

What this means is that for loosely toleranced resistors, one lot may have resistors taken largely from one end of the bimodal distribution, and a later lot may have resistors taken largely from the other end of the bimodal distribution. Both lots meet the specification requirements, but their performance will be decidedly different.

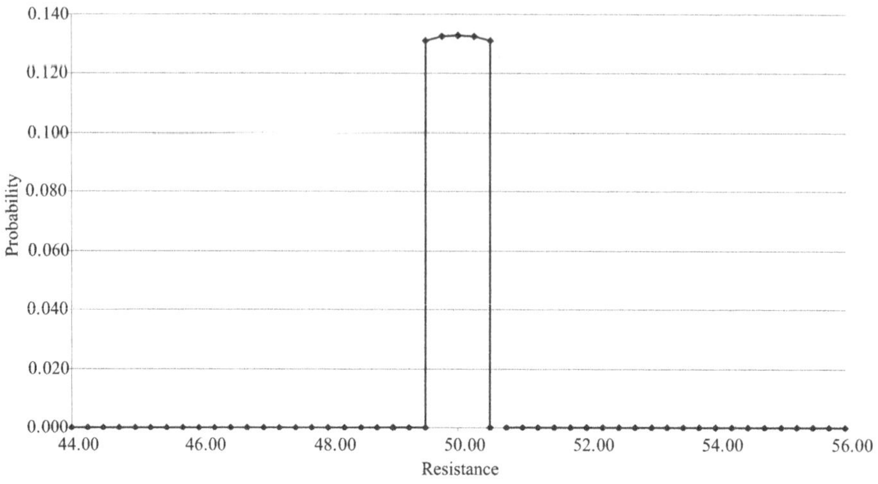

Fig. 10.2 ±1% resistors. After sorting, the distribution at this tight tolerance appears to be nearly uniform.

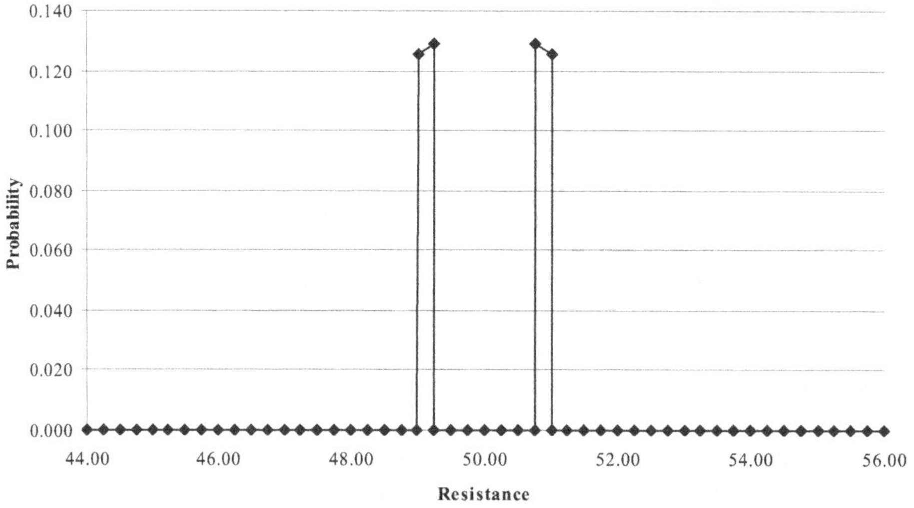

Fig. 10.3 ±2% resistors. After sorting out the ±1% resistors, the distribution becomes bimodal. For less tightly toleranced resistors, the bimodal effect is even more pronounced.

The failure analysis team should understand suppliers' processes in order to identify situations such as the preceding.

Aging. In some cases, the age of a material may make a difference. Many materials have a shelf life defined by their manufacturer. The failure analysis team should consider if any of the hypothesized failure causes can be induced by aging. For example, elastomeric materials may degrade over time. The degradation may be accelerated by heat or atmospheric contaminants. Many materials are hygroscopic (absorb water in the air); such materials may be influenced both by the environment and by aging. A manufacturer may have information on the age of the materials it produces, but it probably will not have such information on materials provided by suppliers. This is another area in which the failure analysis team's purchasing representative can help.

Supplier Changes. As previously noted in this chapter and elsewhere, purchased components, materials, and subassemblies often comprise much of a system. For this reason, the failure analysis team should review all of the areas mentioned earlier with suppliers if the suppliers provide items related to hypothesized failure causes. Sometimes, the purchasing department changes suppliers. Such information is not always readily apparent (even if the company investigating the failure is the one that made the switch in suppliers). Subtier suppliers can also make changes in their suppliers, which is even less visible to the failure analysis team. The failure analysis team's purchasing representative can assist in working with suppliers to identify these changes.

An Explosive Problem

Finding the causes of failures in complex systems is much more challenging than simply examining failed components and determining the reasons why they failed. As explained in this book, during a systems failure analysis the failure analysis team must consider complex system interactions, identify all possible failure causes, and then systematically evaluate each to rule in or out hypothesized failure causes. The cause may be due to a nonconforming component, engineering errors, changes in the process, changes in the design, personnel changes, or other factors. The failure analysis process can be greatly complicated if the failure occurs intermittently, if the failed hardware is not available for examination, or if materials from suppliers induce the failure. The following example describes a cluster bomb failure and explains how the failure analysis team found and corrected the failure cause.

The CBU-87/B, produced by Aerojet, is a cluster bomb that has been used with great effect by the U.S. Air Force. Developed in the 1980s, the system is comprised of a dispenser, a dispenser fuse, and 202 submunitions. Combat aircraft deliver and release the cluster bomb at altitudes and speeds designed to control the submunition pattern on the ground.

After release from the aircraft, dispenser fins deploy immediately to stabilize the device. The cluster bomb free-falls for a specified time, at which point an explosive bolt fires in the dispenser tail section. When this occurs, the dispenser fins tilt, and the dispenser spins up to a preselected rotational speed. When the dispenser attains the required rotational speed, an inertial sensor sends a signal to fire linear charges. The linear charges open the dispenser by peeling away its skin, releasing the submunitions into the airstream. Figure 10.4 shows the operational sequence.

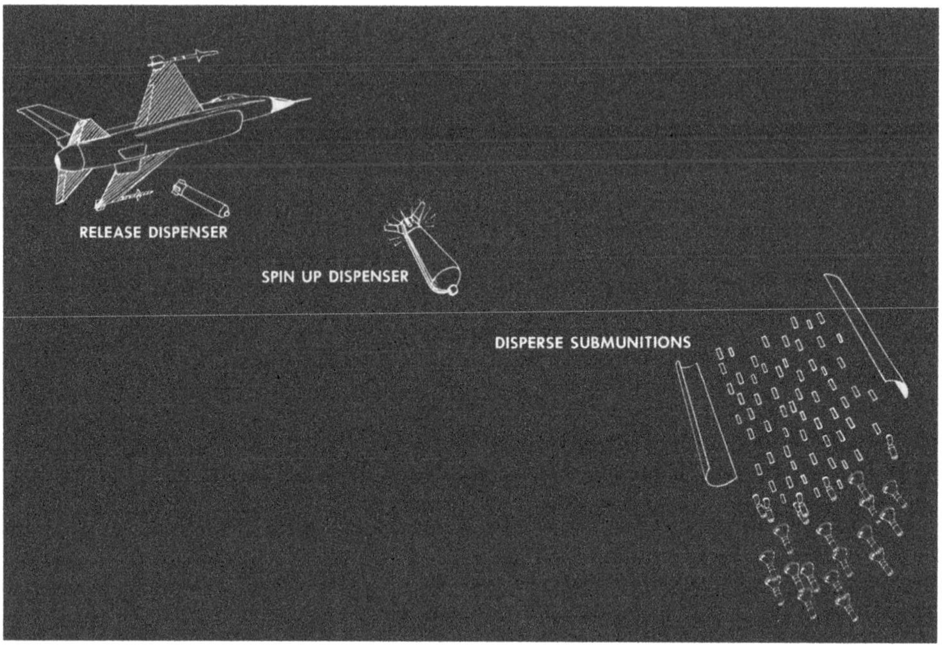

Fig. 10.4 CBU-87/B cluster bomb operational sequence. The aircraft releases the dispenser, which spins up in the airstream. At a preselected dispenser rotational speed, linear charges fire to open the dispenser. This releases submunitions into the airstream. Selecting aircraft altitude, the time the dispenser starts to rotate, and the dispenser rotational speed controls the submunition pattern on the ground.

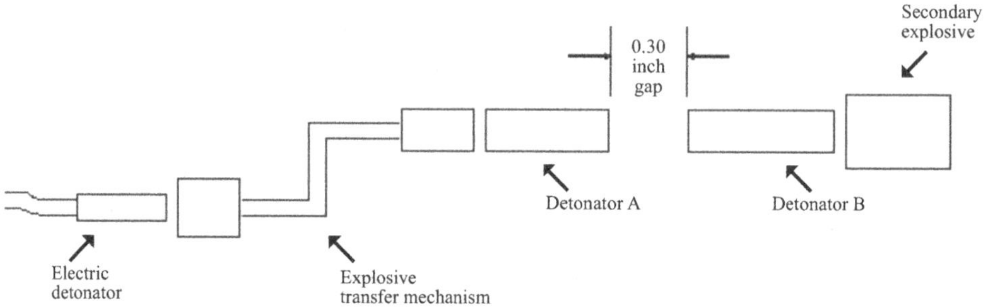

Fig. 10.5 Detonator fuse explosive train. The electric detonator ignites an explosive transfer mechanism, which fires detonator A. This fires across a 0.30 in. gap to ignite detonator B, which then ignites the linear charges.

Firing of the linear charges is controlled by a dispenser fuse with an explosive train, as shown in Fig. 10.5.

The signal to open the dispenser fires an electric detonator into an explosive transfer mechanism. The explosive transfer mechanism ignites detonator A, which fires across a 0.30 in. gap to ignite detonator B. Detonator B then ignites the secondary explosive (the linear charges that open the dispenser). When the dispenser fuse is not armed, a 0.30 in. steel barrier is inserted between detonators A and B. If the electric detonator, explosive transfer mechanism, or detonator A fires inadvertently, the explosive event stops at the steel barrier, and the dispenser does not open. When the device is armed, the steel barrier is removed to allow the explosive event to continue across the gap.

Aerojet built the dispenser fuse to a U.S. government design. The company flight-tested and dropped two dispensers each month for lot-acceptance testing. After beginning initial production, Aerojet experienced no dispenser failures during the first several months of lot-acceptance testing. Then, during one of the monthly flight lot-acceptance tests, one of the dispensers failed to open.

After preparing a fault-tree analysis, the failure analysis team examined components from the recovered (and unexploded) cluster bomb dispenser. The returned components had been damaged too severely to determine the failure cause, so the failure analysis team tested several additional dispenser fuses and linear charges at its test facility. In the test lab, the explosive propagation usually (but not always) transferred across the 0.30 in. gap. The failure analysis team isolated the point of failure to the 0.30 in. interface between detonators A and B. The failure analysis team found that detonator A always fired, but the explosive event did not always transfer across the 0.30 in. gap to ignite detonator B. The failure analysis team tested several dispenser fuses and observed that this failure occurred on an intermittent basis.

The failure analysis team checked all dispenser fuse components against drawing requirements, but they found no nonconformances. The failure analysis team checked the pedigree of the detonators and found that everything was in order. The individual detonators had been acceptance tested by selecting 32 detonators from each lot and firing them into a steel witness block, with the detonator held directly against the witness block. The detonators had always worked, leaving a significant dent in the witness block.

The failure analysis team looked for differences related to the failure causes hypothesized by the fault-tree analysis and listed in the FMA&A. The same explosive component supplier had provided the detonators for years, and, in fact, the supplier had designed both detonators A and B. The only difference the failure analysis team could find was that the recent detonators were from a new production lot. However, the failure analysis team could find nothing about the new lot that was different from prior lots.

The failure analysis team's purchasing member called the detonator supplier. The detonator supplier suggested the Aerojet team visit the detonator manufacturing facility. Aerojet sent two engineers, who met with the supplier's technical staff and toured the factory. There, they learned that the detonator design had an internal steel sleeve, an explosive mix, a thin-walled aluminum cup, and a cap. The thin-walled aluminum container was crimped over the cap, as shown in Fig. 10.6.

The engineers from both companies reviewed all documentation associated with the design, the inspection data, and the process. No anomalies or changes appeared. During the discussion, however, the detonator supplier asked the Aerojet engineers about the explosive transfer mechanism across the 0.30 in. gap. Neither engineer knew the answer to this question. Both assumed heat and shock completed the transfer.

The detonator supplier explained that this particular design relied on a "hurled plate" explosive transfer mechanism (Fig. 10.7). When detonator A fired, it was supposed to create a flat plate at its output end, which would then fly through space and slap into the adjacent detonator. The supplier explained that the kinetic energy associated with this impact was what continued the explosive transfer.

The cluster bomb engineers found this interesting, but they explained to the detonator supplier that it did not explain why the failures occurred. The detonator supplier recommended talking to the production operators who actually manufactured the detonator. The visiting engineers agreed, and all three people visited the production area.

In discussing the problem with the production technician, the engineers learned that the technician crimped the aluminum cup with the detonator output end resting against a rubber block (Fig. 10.8). The engineers asked

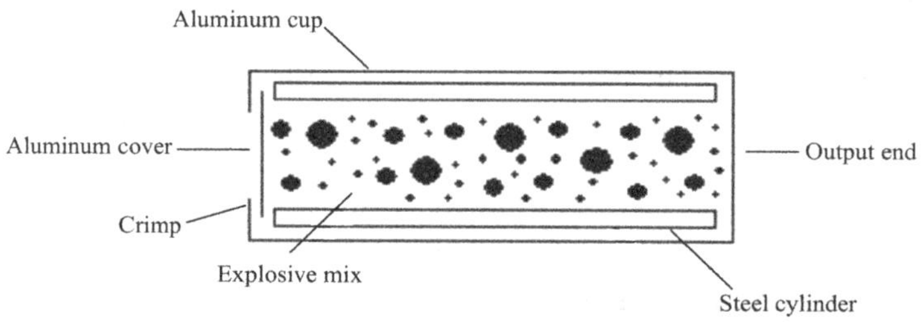

Fig. 10.6 Detonator A design details. The detonator consisted of four components. A steel sleeve formed a barrel inside the aluminum cup. An aluminum cover crimped inside the cup sealed the explosive mix inside the steel barrel.

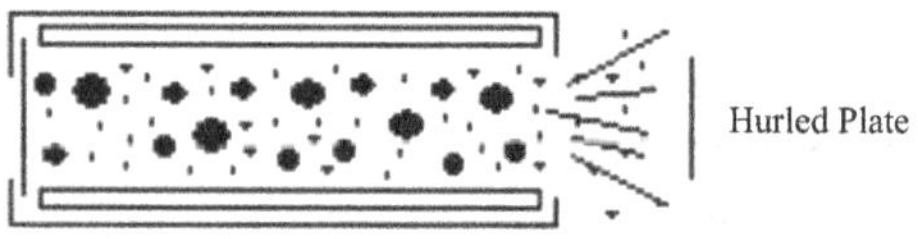

Fig. 10.7 Hurled plate formation. When the detonator fires, it shears the aluminum cup at its output end, creating a plate that is hurled across a gap. When this plate strikes the next explosive train component, kinetic energy ignites it.

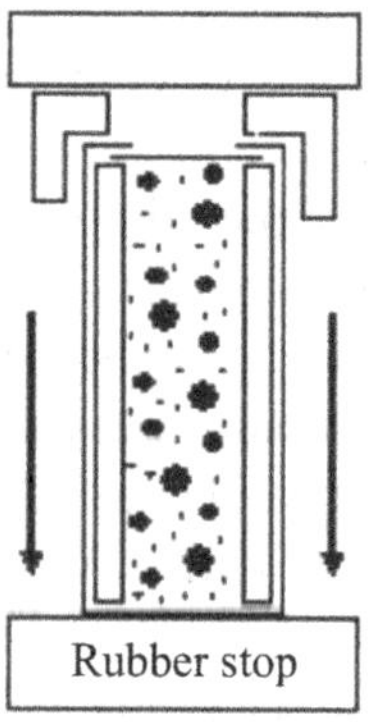

Fig. 10.8 Detonator crimping operation. The aluminum cover at the top of the detonator was crimped in place, with the output end of the detonator held against a rubber stop.

if the technician had made any changes to the process, and the technician explained that approximately two months earlier, he had increased the crimping pressure to obtain a more uniform crimp.

The detonator engineer immediately recognized the significance of this change, and he examined the ends of recently crimped detonators. The output ends displayed varying degrees of concavity induced by the rubber stop. The detonator engineer noted that a concave surface would not form a flat plate but instead deformed into a molten aluminum jet when the detonator fired. In effect, the detonator output end became a small, shaped charge instead of a flat, hurled plate.

The three engineers examined the detonator drawing and found that it had no flatness requirement. (In other words, the drawing did not prohibit the concave condition.) While still at the detonator supplier, the engineers tested detonator outputs, but instead of placing the detonator directly against a witness block when it fired (as was the practice during detonator lot-acceptance testing at Aerojet), the engineers left a 0.30 in. gap to duplicate the dispenser fuse design. The engineers found that detonators with less than 0.05 in. concavity formed a clean plate, which left a circular impression on the witness block. Detonators with more than 0.05 in. concavity left jagged holes in the witness block, clearly indicating that the detonators had not formed the required plate.

Based on this finding, the failure analysis team implemented several corrective actions. The team modified the detonator drawing to specify maximum allowed concavity at the output end. The detonator supplier replaced the rubber stop used during the crimping operation with a steel stop. The failure analysis team modified the detonator-acceptance test approach. Instead of placing the detonator against the witness block, the acceptance test was modified to fire the detonator across a 0.30 in. gap. The team modified the acceptance criteria to include a clear, circular impact on the witness block. The team recommended that Aerojet inspect all detonators in stock, keeping only those with acceptably low concavity. Finally, the team inspected delivered dispensers by visiting munitions storage depots and replaced any detonators having unacceptable concavity. After the failure analysis team implemented these corrective actions, there were no recurrences of this failure.

The CBU-87/B munition failure analysis illustrates several interesting points:

- An undocumented change to the process caused the failure. There was no requirement to document process changes in this area, because the detonator supplier had not identified this to be a critical performance characteristic in its work instructions.
- The process change was extremely subtle and would not have been discovered had the failure analysis team not worked closely with the detonator supplier.

- The detonators that induced the failures conformed to their drawing requirements. The drawing was inadequate, because it did not control features critical to the successful function of the device.
- Aerojet's detonator-acceptance test was inadequate. It did not evaluate the hurled plate mechanism that the design relied on for successful function.
- The failure analysis team did not initially understand how the system worked. (The Aerojet engineers did not know about the hurled plate transfer mechanism.) It was not until this knowledge became available that the team recognized this as another potential failure cause.
- The failure analysis team would not have discovered the failure cause without visiting the supplier and without seeking input from the supplier's production technicians.

Summary

Failures can be induced by changes to the process, the design, the environment, supplier actions, or other factors. Sometimes, failures are caused by changes that induce process deficiencies, and these changes and their effects may not be immediately apparent. Sometimes, failures are caused by not understanding and controlling critical systems characteristics. Sometimes, failures are caused by acceptance-test approaches that fail to duplicate actual operating conditions. In the example included in this chapter, nearly all of the preceding occurred. The failure analysis team resolved the failure by systematically identifying all potential failure causes, involving the supplier, and working to evaluate each. In the process, the failure analysis team uncovered previously unknown design characteristics and identified new failure causes.

The failure analysis team should recognize that changes are constantly made to designs, processes, suppliers, and other areas. The team will discover changes, and these must be evaluated against the potential failure causes hypothesized by the fault-tree analysis.

Example for Group Discussion. Organizations may switch suppliers, processes, people, and designs for cost, quality, delivery, or other reasons. What are the risks associated with these types of changes, and how can an organization manage this risk?

SELECTED REFERENCES

- *Guidelines for Failure Mode and Effect Analysis for Automotive, Aerospace, and General Manufacturing Industries,* Dyadem Press, 2003
- H.M. Tawancy, A. Ul Hamid, and N.M. Abbas, *Practical Engineering Failure Analysis,* Marcel Dekker, 2004

CHAPTER 11

Analytical Equipment

AFTER THE FAULT-TREE, a failure-cause identification method has identified potential failure causes and the failure analysis team has prepared a failure mode assessment and assignment (FMA&A). The team knows specifically what to search for when examining components and subassemblies from the failed system. In this phase of the analysis, several tools and instruments are available to support the examination and analysis of these items. This chapter reviews these methods.

There are numerous techniques and technologies available for examining and analyzing components and subassemblies, which are categorized as follows:

- Optical approaches for visually assessing suspected items under both normal and magnified conditions. Techniques in this area include visual examination, low-power magnification, optical and scanning electron microscopy, and photography.
- Dimensional inspection and related approaches for assessing suspected item dimensions and conformance to drawing and specification requirements and determining if such dimensions are biased within the tolerance band. Techniques in this area include standard measuring equipment, optical comparators, coordinate-measuring machines, surface-finish gages, scales, and torque wrenches.
- Nondestructive test approaches for evaluating internal component positions and relationships, leak paths, and other conditions not readily apparent using visual or other assessment techniques. Techniques in this area include x-ray and n-ray radiography, dye penetrant inspection, fluorescent dye pressure testing, and magnafluxing.
- Mechanical and environmental approaches for evaluating material strength, susceptibility to temperature and humidity cycles, and other related characteristics. Techniques in this area include stress analysis as well as testing for hardness, materials strength, fatigue, environment, and accelerated aging.

- Chemical and composition analysis for assessing material characteristics. Techniques in this area include energy-dispersive analysis of x-rays, spectroscopy, chromatography, and others.

The failure analysis team should select appropriate examination techniques based on hypothesized failure modes. If the team suspects, for example, that cracks in a component may have caused the failure, they should consider techniques that reveal the presence of cracks, crack progression mechanisms, and crack locations. If the team suspects that contaminants could have caused the failure, they should consider techniques that reveal the presence of contaminants. All of the aforementioned approaches are reviewed in this chapter. Appendix A provides a summary of the analytical techniques and their applications.

Optical Approaches

In many cases, the fault-tree analysis hypothesizes component conditions that could have induced the failure. These may include mechanical failures, electrical or electronic component failures, electrical arcing, the presence of contaminants, interference fits, poor welds or solder joints, porosity, cuts, scrapes, witness marks, and other conditions.

In many cases, hypothesized failure causes may be visible to the naked eye. In these situations, a simple visual examination can reveal the presence of cracks, mechanical disconnects or disbonds, leaks, and other failure causes. For example, Fig. 11.1, shows that a capacitor mounted on a circuit card failed. The defective capacitor is easily discernible without magnification.

In many other cases, hypothesized failure causes may not be visible to the naked eye. In these situations, magnification is used to determine if the hypothesized conditions are present. A simple magnifying glass can easily enlarge an area, and, in many situations, low-power magnification may be all that is necessary to show the presence of a hypothesized failure mode. This is often the case for fractured parts, because low-power magnification can be sufficient to reveal surface fractography and, based on this, the fracture mechanism. (Surface fractography is discussed in more detail in Chapter 12.)

Microscopy. In many instances, failure causes cannot be seen with the unaided eye or under low-power magnification. For example, this may occur when looking for minute contaminants, metallurgical inclusions, material voids, fractures too tiny to see under low magnification, or other items too small for the aforementioned examination methods. In these cases, microscopic examination is required.

Optical microscopes offer magnifications of 10 to as much as 2000 times the actual size. They are generally easy to use, with very little or no training required. Most microscopes can be used in conjunction with a

Fig. 11.1 Failed capacitor. The defective capacitor can easily be seen without magnification. Sometimes, a simple visual inspection can reveal the presence of a hypothesized failure cause.

camera to permanently record images. Microscope costs can range from hundreds to several thousands of dollars. Optical microscopy disadvantages include magnification and specimen-size limitations.

Scanning electron microscopes (SEMs) offer significantly higher magnification, with some going as high as a million times the actual size. In addition to these significant magnification advantages, SEMs are relatively compact. (Most units fit on a desktop.) Figure 11.2 shows a typical SEM.

The SEM sputtercoats the item to be examined with a very thin layer of gold or palladium. An electron source within the SEM then bombards the specimen with electrons, which reflect off of the metal layer. A detector captures the reflected electrons, and a processor converts this information into an image. Most SEMs display the image on a computer screen. The SEM technician can observe the image and preserve it with a digital camera or (on older units) a Polaroid photo for future reference.

The SEM technology has a few disadvantages. An SEM typically costs well over $100,000. Operating an SEM requires specialized training. The SEM technology has significant specimen-size restrictions and cannot be used for viewing organic materials.

The SEMs have been widely available for several decades, and many medium-to-larger-sized organizations have on-site scanning electron microscopy capabilities. Many commercial laboratories offer scanning electron microscopy services at relatively low prices, so if the failure analysis

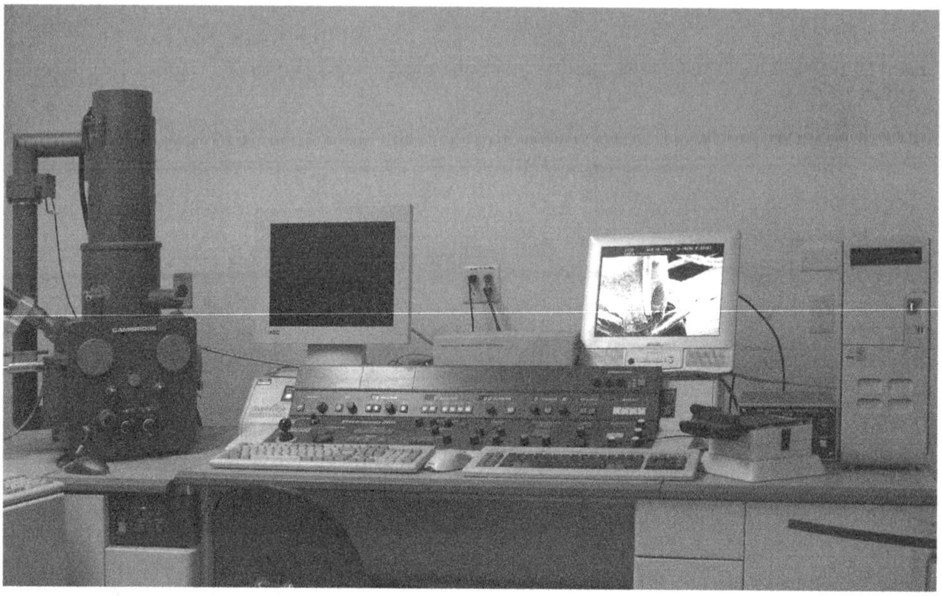

Fig. 11.2 Scanning electron microscope. These analytical systems are usually small enough to fit on a desktop. The specimen chamber is the tall gray cylinder on the left. Courtesy of OCM Laboratories, Anaheim, CA

team's organization does not own an SEM, such services are readily available from outside sources.

Photography. In addition to the optical, low-power magnification, and microscopy approaches described earlier, photographing the system and its components makes sense from several perspectives:

- The photographs become a permanent record of the failure analysis findings (Fig. 11.3, 11.4).
- Photographs of the system, subsystem, and components during each step of the disassembly process can be used for subsequent review. This procedure is especially helpful if additional failure causes are unearthed after the system has been disassembled.
- The photographs can be used in the failure analysis report to illustrate findings and to support the failure analysis team's conclusions and recommendations.
- High-resolution digital images can be enlarged on the computer, offering another way to examine components and assemblies under low-power magnification.

The preferred photographic approach is to use a digital single-lens reflex camera with a tripod and a 50, 105, or 180 mm macro lens. Macro lenses provide sharper images and allow close-up photography, which is usually required in a failure analysis. A tripod ensures a well-focused photograph free of any motion-induced blurring. Currently, digital photography

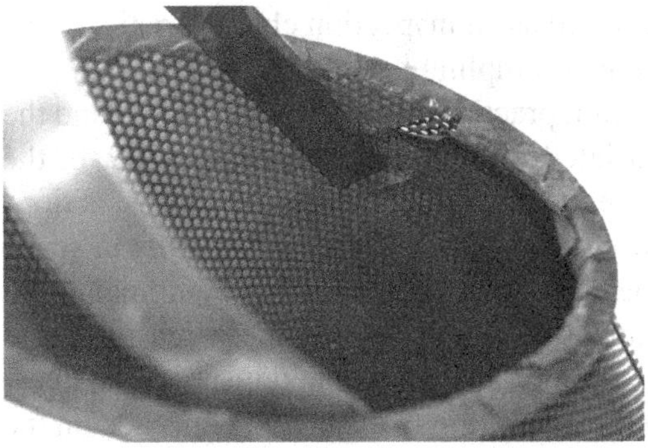

Fig. 11.3 Failed water treatment plant strainer screen that allowed sand to enter the treatment plant, ultimately destroying a $750,000 system. Photos such as these become valuable parts of the failure analysis, because they allow easy understanding of the failure cause and create a permanent record for future reference.

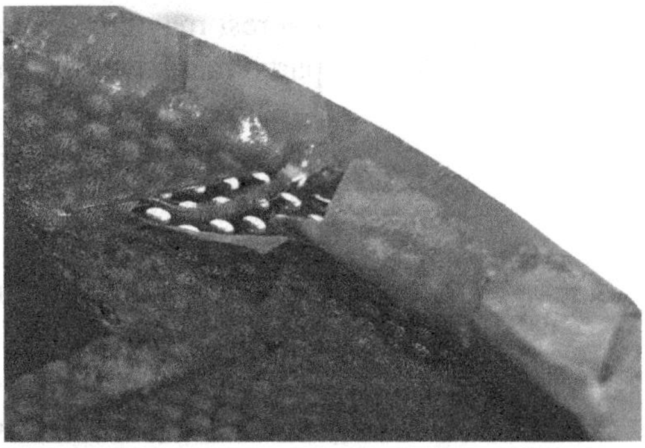

Fig. 11.4 Close-up crop of Fig. 11.3. Digital images with higher resolution allow for close cropping with good detail retention.

is greatly preferred over film photography. (The image is instantly available and can be examined immediately to confirm proper focus, composition, and exposure.) Cameras with larger image sizes (10 megapixels and up) allow the area of interest to be cropped with little or no loss of detail (Fig. 11.4).

Dimensional Inspection and Related Approaches. During the failure analysis, the failure analysis team should hypothesize many failure causes. These hypothesized causes may be present in many different parts, but not all parts of the system contain potential failure causes. Accordingly, the failure analysis team need not inspect every dimension on every part, only those dimensions and characteristics that could have induced the failure.

A good way to communicate which parts require specific inspections is to provide an engineering drawing with the dimensions of interest circled on the drawing. This serves two purposes:

- It identifies areas requiring inspection.
- It identifies if the drawing adequately defines areas required to prevent the failure. For example, in the case of the CBU-87/B fuse failure addressed in Chapter 10, the failure analysis team found that detonator concavity (a critical characteristic) had not been defined on the detonator drawing.

As potential failure-inducing dimensional nonconformances are identified, the failure analysis team may be tempted to rule these out, based on the parts having passed their prior inspections. This would be a mistake for several reasons. The inspector could have performed the inspection incorrectly, the inspection device could be out of calibration, the dimension of interest may not have been identified as an inspection characteristic, or the parts may have been inspected on a sampling basis.

Those unfamiliar with inspection practices may not fully understand the complexities and nuances associated with some inspections. Many of the inspection tools described subsequently are complex. Many inspections require trigonometric calculations, computing theoretical points in space from which to take measurements, setting up measuring equipment, and performing complex mathematical calculations. For these reasons, it is generally a good idea to witness failure-analysis-related inspections, question assumptions made during the inspection to ensure concurrence on the approach, and understand what the inspection is actually measuring. In most cases, this simply educates the observer. In some cases, the failure analysis team may find that the inspection was performed incorrectly (and, perhaps, that the incorrect inspection technique allowed nonconforming items to escape), or they may discover that the inspection equipment is overdue for calibration. In some cases, the failure analysis team may find that the drawing requires revision to correct poorly defined dimensions or missing dimensions.

The failure analysis team's quality-assurance representative can coordinate the use of various types of inspection equipment to determine if hypothesized nonconformances are present. Typical inspection equipment includes:

- *Standard measuring equipment:* Rules, calipers, micrometers, gage blocks, feeler gages, sine plates, protractors, pin gages, and thread gages are used to evaluate dimensional compliance.
- *Optical comparator:* An optical comparator is a projection system that projects the outline of a part onto a screen marked with an *x-y* and radial grid. By comparing the shadow of the part being measured to the grid, inspectors determine if the part meets requirements.
- *Coordinate-measuring machine:* The coordinate-measuring machine uses a probe linked to a computer to determine positions in three-dimensional space. By touching the part in various locations with the probe, the coordinate-measuring machine can define the part dimensions.
- *Surface-finish gages:* These devices assess surface finish. There are two types. The more subjective of the two consists of a set of standards with various surface finishes. The inspector compares these to the surface finish being assessed and visually selects the gage surface finish most similar to the surface being evaluated. The more accurate surface-finish gage uses an interferometer. The interferometer measures surface finish by dragging a pin across the surface. The pin deflection is assessed by a microprocessor, and the gage provides a direct surface-finish readout.
- *Scales* are used to measure force (e.g., the force exerted by a spring).

- *Torque wrenches* are typically used to install threaded fasteners to their specified torques. Torque wrenches can also be used to provide a rough assessment of the torque to which a threaded fastener was installed.

Nondestructive Testing

Standard inspection techniques, as described in the preceding section, are nondestructive in that they do not alter or damage the item being inspected. There are also a number of other nondestructive testing techniques for inspection and examination. These techniques are typically used for:

- Inspecting the interior of components or subassemblies (with methods such as ultrasonic inspection, x-ray inspection, and neutron-beam inspection)
- Inspecting component surfaces for leaks (with dye penetrant and pressurized fluorescent dye inspection techniques) or finding cracks (with dye penetrant and magnetic particle inspection)

X-ray and neutron beam methods are examples of inspection techniques to examine the inside of objects to check for the presence of hypothesized failure causes. X-ray imaging techniques radiate an object with short-wavelength electromagnetic radiation. This allows one to view internal components, cracks, porosity, and other things not visible from outside the object (Fig. 11.5). An x-ray directs radiation through an object, with film placed on the other side. The image is captured on the film, which is subsequently viewed as a negative. X-rays can be used as a normal part of an inspection process (for example, to detect cracks, weld porosity, casting porosity, and other internal defects), for showing the relative position of internal components, or for other failure analysis purposes.

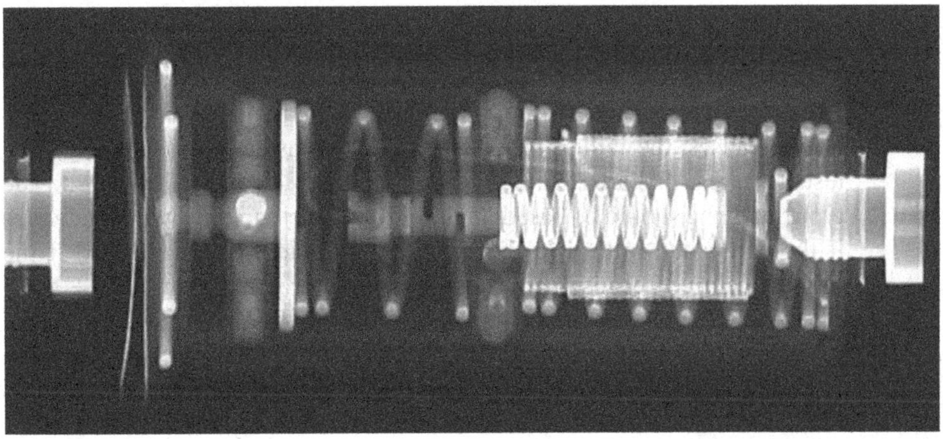

Fig. 11.5 X-ray image showing the internal components of a complex initiator and the component positions

The use of neutrons is another approach that allows one to see inside an object. Neutron radiography is more specialized and has more capabilities than x-ray radiography. Unlike conventional x-ray radiography, neutron beams pick up details that x-rays do not. For example, n-rays show the presence of plastics, oils, and water, while x-rays do not.

Dye penetrant inspection is used for finding cracks that may not otherwise be visible to the naked eye. The technique involves immersing the suspect component in a fluorescent liquid dye for several seconds, removing it, and then wiping the component clean. After a short time period, the component is then subjected to illumination with a black light in a darkened area. The concept is that the fluorescent dye seeped into the crack, and after the component has been wiped clean, some of the dye (through capillary action) seeps back out of the crack. When exposed to the black light, the dye outlines the crack. This technique is useful for finding cracks in components and for finding leak paths if the leak path extends to the surface of the item being examined.

Pressurized fluorescent die testing is another technique used for finding leak paths. In this approach, the suspect subassembly is immersed in a fluorescent dye. The vessel containing the fluorescent dye and the suspect subassembly is then pressurized. After the pressure is removed, the subassembly is wiped clean from the exterior, disassembled, and then examined under a black light in a darkened area (similar to the approach used for dye penetrant inspection). If a leak path allowed the fluorescent dye to enter the subassembly, the leak path appears under black light illumination.

Magnafluxing is a specialized technique used for finding cracks in ferrous components. Iron filings are sprinkled onto the part in areas where cracks are suspected, and a magnetic field is applied to the part. If a crack is present, the magnetic field is disrupted along the crack, and the iron filings arrange themselves to follow the crack path. When this occurs, crack presence and location can be confirmed.

Mechanical and Environmental Testing

If the mechanical properties of a component are hypothesized as potential failure causes, the failure analysis team can consider various mechanical tests to assess this area. This is typically done if a part has fractured. The type of testing should be based on the surface fractography, which indicates if the part failed in tension, bending, compression, or fatigue, and if the failure occurred because the material was too hard or too soft. (These fracture modes are discussed in Chapter 12.) Various mechanical tests are used to evaluate each fracture mode.

Perhaps the most common mechanical test is the hardness test. This is of interest in failure analyses because materials may not perform as required if they are outside their specified hardness range. For example, if steel is heat treated to a hardness level that is too high, it may fail in a brittle manner. Similarly, if it is too soft, it may fail in a ductile manner. If the fault tree hypothesizes such situations, it is relatively easy to perform a hardness test to determine if the material is of the specified hardness.

The approach for testing hardness is straightforward and is called a Rockwell hardness test. A hardened or diamond-tipped punch is pressed against the specimen surface with a known force to form a dimple. Softer materials have bigger dimples than harder specimens. The depth and diameter of the dimple formed in this operation corresponds to the specimen hardness.

Many companies have materials hardness-testing equipment on site. If the failure analysis team finds that hardness-testing capabilities are unavailable in-house, the services are readily available in any major city at a relatively low cost.

Load testing is another method used for evaluating materials. In this approach, a sample of the material is cut to a standard tensile-testing configuration and subjected to an increasing tensile load. As the load increases, the technician records elongation data. The loads are converted to stress (by dividing the load by the cross-sectional area of the standard tensile test specimen), and the elongations are converted to strain (by dividing the elongation by the original length). The technician then prepares a stress-strain diagram. This can be compared to the stress-strain diagram for the specified material, which allows the failure analysis team to determine if the material meets its specification requirements.

A more direct approach is to simply subject identical suspect components to the worst-case loads expected to be experienced in service. If the component fails in a manner hypothesized as causing the failure at or below these loads, the failure analysis team knows that the parts are either defective or underdesigned. These assessments are governed by the distributions of in-service loads and component strengths, a topic covered in more detail elsewhere in this book.

The testing described earlier can be supported by stress analysis, either prepared manually or by using more sophisticated computer modeling techniques (such as NASTRAN). In many cases, however, analytical methods may not be as conclusive as the test approaches outlined previously.

If the component fractured in a manner indicating a fatigue failure, the part can be tested for its resistance to fatigue loading. Most frequently, the hardness test is used for this assessment. In some cases, the failure analysis team may wish to subject the part to repetitive stress fluctuations to measure its life against varying levels of stress fluctuation, although this is a fairly specialized and expensive test. In most cases, if the part failed due to fatigue and the material meets its specification requirements, the failure analysis team may safely conclude that the part is underdesigned for its intended loads.

Components that fail in fatigue may have had stress fluctuations induced by vibration. If the failure analysis team suspects this is the case, similar components can be subjected to suspected worst-case vibratory loads. Most large cities have commercial test laboratories that offer vibration-testing services.

The failure analysis team may encounter situations in which hypothesized failure causes are induced by aging, repetitive temperature cycling, or exposure to humidity, ultraviolet radiation, or high or low temperatures. Most commercial testing facilities offer environmental and accelerated aging tests to evaluate these hypothesized failure causes.

Materials Analysis

Failures are often caused by contaminants or by the use of incorrect materials. If the hypothesized failure causes include incorrect materials or contamination, several technologies are available for evaluating suspect materials. Some manufacturers have in-house materials identification capabilities. If these capabilities are not available to the failure analysis team, many commercial laboratories can provide the services. Chemical and materials analysis technologies include energy-dispersive analysis of x-rays, spectrographic techniques (including electron probe microanalysis, Fourier transform infrared spectroscopy, mass spectrometry, secondary ion mass spectrometry, and Auger analysis), and chromatography. These technologies are described as follows.

Energy-Dispersive Analysis of X-Rays. Perhaps the most common materials analysis technology is energy-dispersive analysis of x-rays (EDAX). The EDAX equipment is commonly bundled with SEM equipment (Fig. 11.6). It can also be purchased as an aftermarket accessory to existing SEM systems.

When an SEM bombards a specimen with electrons, the specimen emits x-ray energy in addition to the reflected electrons. The x-ray energy has characteristics that can be correlated with various elements. Capturing and processing the emitted x-rays allows one to determine which elements are present. The failure analysis team can use EDAX technology to determine if contaminants are present and, in some cases, if the component has been fabricated from the correct material.

The typical EDAX output is an *x-y* plot, with peaks corresponding to various elements. Because EDAX technology identifies elements and not compounds, analysts who use this technology must make inferences about the compounds present based on element identification.

The EDAX technology offers a relatively inexpensive approach for element identification. Specimens have the same size limitations as an SEM. (The same specimen-mounting area is used by the EDAX system.) The EDAX operation requires a trained technician. It cannot be used on organic specimens, and the equipment can be expensive to maintain.

Spectrometry uses light to identify the presence of various elements and materials. The concept is that light is either absorbed or emitted by

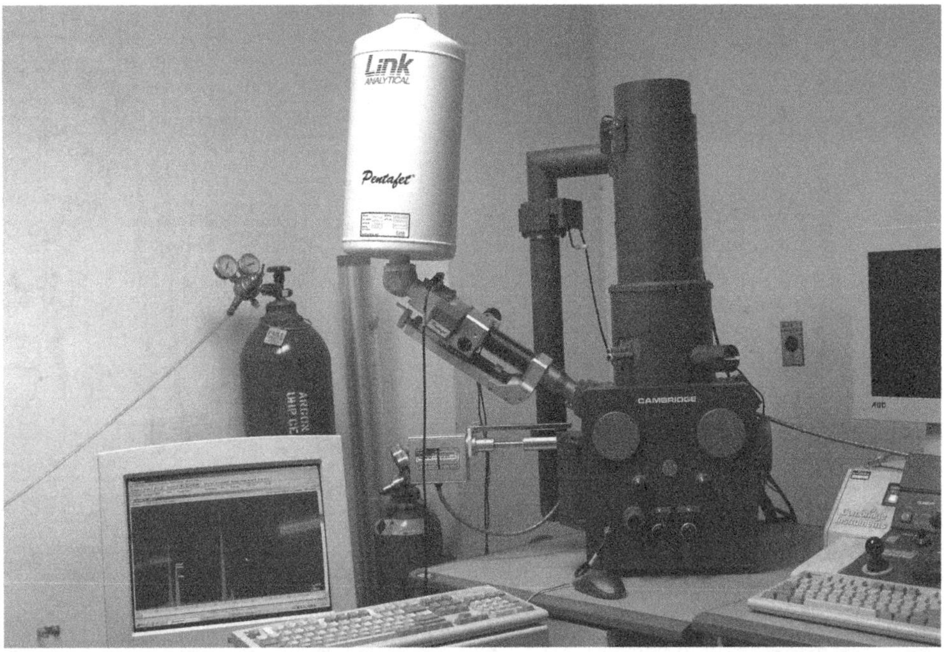

Fig. 11.6 Energy-dispersive analysis of x-rays system connected to scanning electron microscope shown in Fig. 11.2. Note the display on the left, showing elemental peaks. Courtesy of OCM Laboratories, Anaheim, CA

various materials. Captured light can be correlated to different materials. Several materials analysis technologies rely on spectrometry.

Electron probe microanalysis (EPMA) is a materials analysis technique that can be bundled with an SEM. Instead of relying on x-ray energy (as in the EDAX approach), EPMA uses a wavelength-dispersive spectrometer that is approximately an order of magnitude better than EDAX at detecting elemental peaks. That fact notwithstanding, EDAX is still much more commonly used than EPMA.

Fourier transform infrared (FTIR) spectroscopy is another technology used for element and compound identification. Figure 11.7 shows an FTIR system. In the failure analysis context, FTIR technology is most frequently used for contaminant identification, although it can also determine if the correct material was used. The FTIR procedure typically involves washing a specimen and capturing the solution used for the wash, drying it on a slide, and bombarding the dried solution with photons in the infrared range. Some FTIR systems also can mount scrapings from the material being investigated and analyze the scrapings directly.

The FTIR technology uses a detector that captures infrared energy from the specimen as it is being bombarded with photons. A processor with regression software correlates the captured infrared energy to vibration levels associated with interatomic bonds of various elements. The FTIR system provides an *x-y* plot similar to that of the EDAX system. The *x-y* plot shows

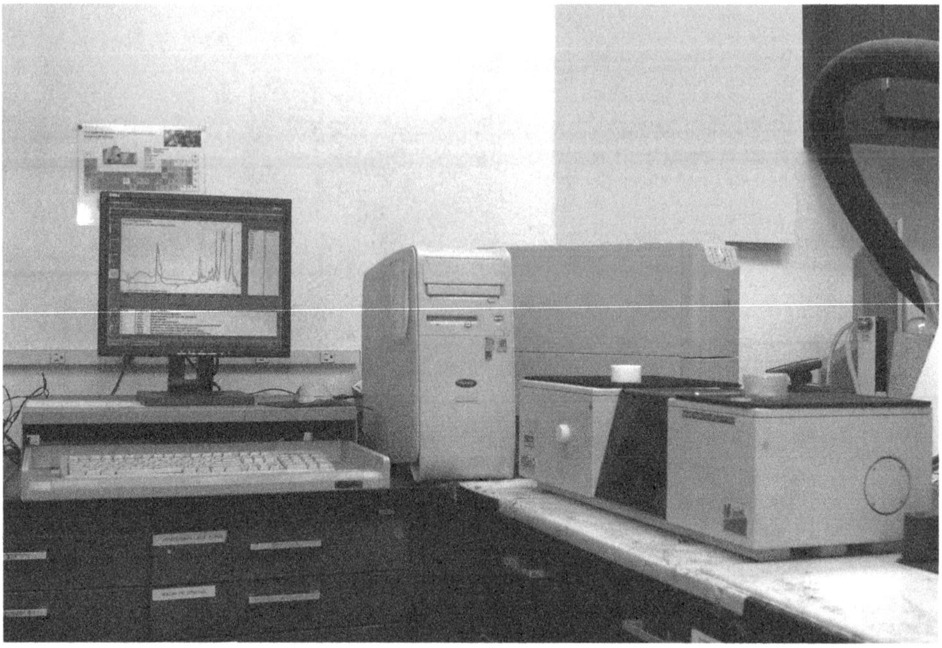

Fig. 11.7 Fourier transform infrared spectroscopy equipment. The apparatus on the right houses and illuminates the sample. A computer houses algorithms and data to which the trace is compared. The monitor on the left displays the compound trace and a suggested material comparison trace.

peaks that correspond to various elements. The EDAX system typically has a library of known compounds and can display these along with the sample trace, expediting compound identification.

Similar to EDAX, FTIR technology identifies elements and not compounds; however, the modern FTIR system has a large library of known compound traces. When the FTIR system has identified the sample compound elements and displayed their *x-y* trace, it then suggests similar traces from its library on the same display, allowing the analyst to identify the compound.

Mass spectrometry is another technology used for materials identification. In a mass spectrometer, ions bombard the sample. The ion bombardment reduces the sample molecules to ionized fragments. These fragments are then subjected to magnetic and electronic fields, which segregate the ions by mass and by charge. A detector measures both the mass and the charge. The mass spectrometer processes this information using algorithms that can correlate the mass-to-charge ratio to various compounds.

Secondary ion mass spectrometry (SIMS) is a specialty elemental analysis technology widely used by the semiconductor industry. SIMS is useful for identifying trace amounts of elements, especially where high purity is required. SIMS can detect elements at much lower concentrations than the other materials analysis technologies described previously.

SIMS technology involves bombarding a small specimen with ions or laser energy to free ions from the specimen. A mass spectrometer then identifies the elements that are present. SIMS technology can also identify element concentrations. Because the bombardment process actually bores a hole in the specimen at the bombardment site, SIMS technology can identify element presence and concentration versus depth from the surface.

SIMS systems are quite expensive to own and maintain; as a consequence, very few organizations use these systems. SIMS technology requires a search for suspected elements. (The technology does not provide the same broad-based search capabilities as the other technologies described previously.) SIMS technology does not work with insulating materials.

Auger technology (pronounced "oh zhay") is an extremely specialized elemental analysis approach. Auger systems are used for analyzing contaminants on the surface or in very thin layers near the surface. In an Auger system, the specimen is illuminated with an electron stream (similar to the approach used by scanning electron microscopy). The specimen then emits Auger electrons, which are electrons emitted by the excitation and de-excitation of atoms. These Auger electrons can be captured by a detector and correlated to various elements. Not many Auger systems exist. They are large (the system fills an entire room) and expensive. The service is generally not available from commercial laboratories.

Chromatography refers to processes that separate compounds for subsequent chemical analysis using refraction. The process consists of stationary and mobile phases, and it can be accomplished with the material in either a gaseous or liquid state. The mobile phase moves through the stationary phase, with various constituents in the mobile phase moving through the stationary phase at different rates. This movement at different rates allows the equipment to separate the constituents. As these various constituents emerge from the stationary phase, they are analyzed using refractive techniques to identify the chemical composition. Older chromatography technology moved the mobile phase through the stationary phase via capillary action. More modern equipment moves the mobile phase through the stationary phase with a pump. These are called high-performance liquid chromatography systems (Fig. 11.8).

Chromatography output is similar to that of the other materials analysis technologies. The system provides an *x-y* plot showing the compounds that are present. The analyst makes inferences about the material based on the *x-y* plot. The chromatography system also provides suggested trace overlays to assist in the materials identification process, as described previously for FTIR technology.

For failure analysis purposes, chromatography can be used to determine if components related to hypothesized failure causes are the correct material and to identify suspected contaminants.

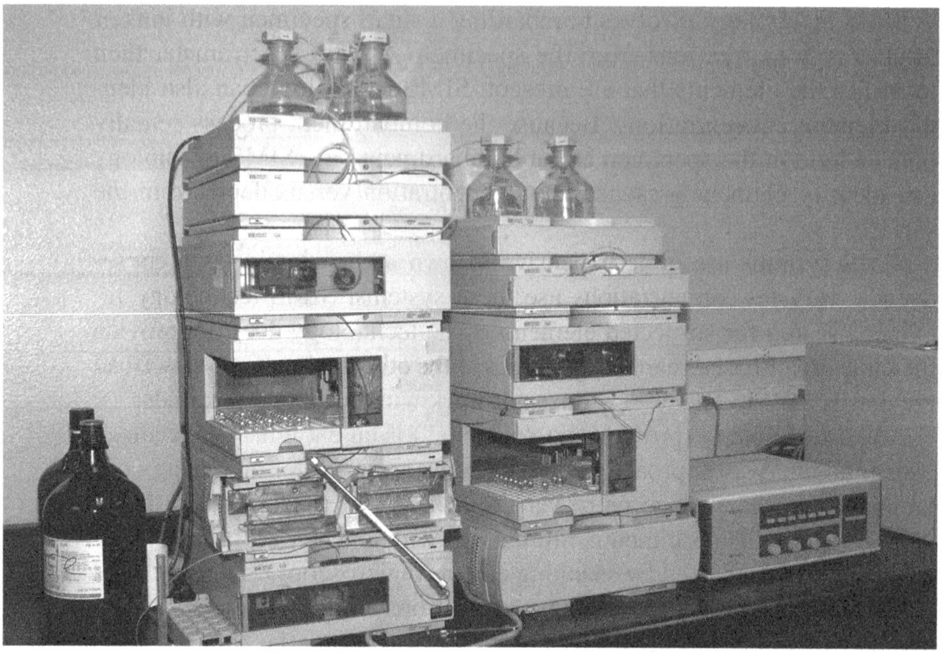

Fig. 11.8 High-performance liquid chromatography equipment pumps the mobile liquid phase through the stationary phase and identifies emerging compounds using refractive techniques.

Summary

Several tools and instruments are available for examining and analyzing suspect components and subassemblies after the failure analysis team has prepared a fault-tree analysis and an FMA&A. Standard measuring equipment allows the failure analysis team to determine if suspect parts meet dimensional requirements. Magnification (using a magnifying glass, optical microscopy, or scanning electron microscopy) allows the failure analysis team to determine if witness marks, cracks, porosity, foreign materials, and other characteristics are present. Materials analysis equipment (EDAX, spectroscopy, and chromatography) allows the failure analysis team to identify contaminants and to determine if components were fabricated from the correct materials. Hardness-testing equipment allows the failure analysis team to determine if the material was correctly heat treated. Radiographic techniques (x-ray and n-ray) allow the failure analysis team to see inside subassemblies prior to disassembly to determine if all parts are present and in the correct position.

Appendix A provides a quick reference showing how the aforementioned analysis tools can be used to evaluate various failure causes.

Example for Group Discussion. Should the technologies discussed in this chapter be used during the course of a failure analysis to verify that all components have been constructed using the proper materials and that these components meet all dimensional requirements? In other words, should the failure analysis team perform a first-article inspection on any or all system components?

Suppose that during the course of a failure analysis, the failure analysis team uses the technologies described in this chapter and finds that the components and subassemblies related to hypothesized failure modes meet their dimensional and material requirements, all components were assembled correctly, and the system was operated in its intended environment. What conclusions should the failure analysis team form if this occurs?

SELECTED REFERENCES

- V.J. Colangelo and F.A. Heiser, *Analysis of Metallurgical Failures,* John Wiley and Sons, 1987
- I. Le May, Examination of Damage and Material Evaluation, *Failure Analysis and Prevention,* Vol 11, *ASM Handbook,* ASM International, 2002
- L.E. Murr, *What Every Engineer Should Know about Material and Component Failure, Failure Analysis, and Litigation,* Marcel Dekker, 1987
- H.M. Tawancy, A. Ul Hamid, and N.M. Abbas, *Practical Engineering Failure Analysis,* Marcel Dekker, 2004
- D.J. Wulpi, *Understanding How Components Fail,* ASM International, 2000

CHAPTER 12

Mechanical and Electronic Failures

COMMON FAILURE characteristics exhibited by mechanical and electrical components are the focus of this chapter. Before delving into how to analyze failed hardware, the topic must be considered from two perspectives:

- One possibility is that the system failed because parts were nonconforming to drawing requirements, they were assembled improperly, contaminants were present, or the system operated outside its intended environment. In this case, the design was not at fault.
- Another possibility is that the system failed even though all parts in the system met their drawing requirements, the system was properly assembled, no contaminants were present, and the system operated inside its intended environment. If this is true, the system design is inadequate. Chapter 15 discusses how to find these types of problems. For now, the reader should recognize that a review of all of the components in the system may reveal no discrepancies; in that case, the failure cause lies elsewhere.

In the course of analyzing components from the failed system, the failure analysis team will probably find some components that do not meet drawing requirements. No matter how much inspection a manufacturer performs, some nonconforming components will find their way into the product. There are many reasons for this;* for now, the reader should accept that this is the case. Given that this occurs, the failure analysis team must guard against concluding that simply because a part is nonconforming it must have caused or somehow contributed to the failure. As nonconformances are discovered, the failure analysis team should certainly seek

* Studies have consistently shown that for other than automated inspections, human inspectors miss 20% or more of the defects when performing inspections. Also, many items are not inspected on a 100% basis; instead, sampling inspection techniques are used. Sampling can allow defects in the population not inspected.

to correct the conditions that allowed these nonconformances to occur, but the team cannot jump to the conclusion that just because a component is nonconforming it must have caused the failure. Each nonconformance should be compared to the fault-tree analysis and the failure mode assessment and assignment (FMA&A) to determine if the nonconformance is related to the failure being analyzed or if it is simply a coincidental finding.

Having pointed out the previous topics, the discussion turns to how to find information from failed items.

Metallic Component Failures

Metallic components can fracture in various manners. They can corrode, fracture, deform, interfere with one another, disconnect, or leak. They may not meet drawing requirements and fail as a nonconformance consequence. An examination of the failed system and its components can confirm or refute the existence of hypothesized failure modes in each of these categories.

Corrosion

Corrosion is the deterioration of a metal due to an electrochemical reaction with the environment. From a failure analysis perspective, corrosion is of interest from several perspectives. The existence of corrosion may constitute a failure for aesthetic reasons; corrosion may lead to the corroded component fracturing; or corrosion may lead to increased electrical resistance (which could induce degraded performance in electrical or electronic systems).

All corrosion is galvanic in nature. It is induced by electrical reactions between two different metals in the presence of an electrolyte (as occurs in a battery). The extent of the corrosive reaction depends on the dissimilarity of the metals (how far apart they are in the galvanic table) and the nature of the electrolyte. In many cases, it may appear that no electrolyte is present, but humidity, sea spray, adhesives, and so forth may become unintended electrolytes. Insulatory materials (paint or other special coatings) can be used to inhibit or prevent galvanic corrosion; the absence or incorrect application of these coatings should be considered as potential galvanic corrosion causes. From a failure analysis perspective, galvanic corrosion should be considered as a failure cause wherever corrosion, fractures, or increased electrical resistance is suspected. In many cases, galvanic corrosion may not immediately be visible; dissimilar materials may require disassembly to allow assessment of galvanic corrosion presence. MIL STD 889 (*Dissimilar Metals*) is an excellent reference for assessing susceptibility to galvanic corrosion based on materials selection.

The term *uniform corrosion* is used to describe normal corrosion (rust, oxidation, etc.); however, even this corrosion is a form of galvanic corrosion. Uniform corrosion is actually microscopic galvanic corrosion induced by microreactions between tiny amounts of dissimilar materials. For example, stainless steel can exhibit corrosion in the presence of strong electrolytes if microscopic inclusions of other materials are present. Passivation, a chemical etching process designed to remove such inclusions from the surface of stainless steel, prevents this from occurring. For most materials, coatings (e.g., paints or other insulating coatings) are used to prevent corrosion. The concept is that the coating keeps the electrolyte from forming a bridge between the dissimilar metals. From a failure analysis perspective, potential causes of uniform corrosion include inadequate or inappropriate coatings, the presence of strong electrolytes, and discontinuities in the coating.

Stress-corrosion cracking can be induced by a combination of corrosion and tensile stresses. (Tensile stresses are frequently present in metallic components.) If a fracture occurs in a normally ductile material, stress-corrosion cracking should be considered as a potential cause (especially if the environment provides an electrolyte).

Apparent fatigue failures (discussed in more depth later in this chapter) should also be assessed from a corrosion fatigue perspective. Corrosion fatigue failures are induced by a combination of cyclic loading and the presence of an electrolyte. Microscopic examination of the fracture surface differentiates a pure fatigue failure from a corrosion fatigue failure.

Fractures

In addition to the fractures induced by corrosion, fractures in metallic components may be stress induced. Stresses may result in tensile, brittle, compressive, or fatigue failures. These fracture modes are summarized in the fracture mode identification chart in Table 12.1, with additional description as follows.

Whenever fractures occur, the failure analysis team should request a stress analysis to determine if the design is adequate. Stress analysis, mentioned in Chapter 11, identifies the stresses in the member under anticipated worst-case operating loads and compares these stresses to the strength of the material. The analysis also considers the effects of stress raisers in the component design. Such things as sharp corners or other dimensional characteristics that greatly increase stresses experienced by a component can induce stress raisers.

Stress analysis should also consider the distribution loads (both the mean and standard deviation) to assess the probability of the load exceeding the component strength. In addition, dimensional inspections should be performed to determine if the failed item met its drawing requirements. (If the part is undersized, the stress increases.) Corrective actions may involve design changes to strengthen the part (either by making it larger or by eliminating stress raisers), process changes (if the part is found to be nonconforming), or selection of a different material with greater strength.

Table 12.1 Fracture mode identification chart

Method	Instantaneous failure mode(a)		Progressive failure mode(b)			
	Ductile overload	Brittle overload	Fatigue	Corrosion	Wear	Creep
Visual, 1 to 50 × (fracture surface)	• Necking or distortion in direction consistent with applied loads • Dull, fibrous fracture • Shear lips	• Little or no distortion • Flat fracture • Bright or coarse texture, crystalline, grainy • Rays or chevrons point to origin	• Flat progressive zone with beach marks • Overload zone consistent with applied loading direction • Ratchet marks where origins join	• General wastage, roughening, pitting, or trenching • Stress-corrosion and hydrogen damage may create multiple cracks that appear brittle	• Gouging, abrasion, polishing, or erosion • Galling or storing in direction of motion • Roughened areas with compacted powdered debris (fretting) • Smooth gradual transitions in wastage	• Multiple brittle-appearing fissures • External surface and internal fissures contain reaction-scale coatings • Fracture after limited dimensional change
Scanning electron microscopy, 20 to 10,000 × (fracture surface)	• Microvoids (dimples) elongated in direction of loading • Single crack with no branching • Surface slip band emergence	• Cleavage or intergranular fracture • Origin area may contain an imperfection or stress concentrator	• Progressive zone: worn appearance, flat, may show striations at magnifications above 500 × • Overload zone: may be either ductile or brittle	• Path of penetration may be irregular, intergranular, or a selective phase attacked • EDS(c) may help identify corrodent	• Wear debris and/or abrasive can be characterized as to morphology and composition • Rolling contact fatigue appears like wear in early stages	• Multiple intergranular fissures covered with reaction scale • Grain faces may show porosity
Metallographic inspection, 50 to 1000 × (cross section)	• Grain distortion and flow near fracture • Irregular, transgranular fracture	• Little distortion evident • Intergranular or transgranular • May relate to notches at surface or brittle phases internally	• Progressive zone: usually transgranular with little apparent distortion • Overload zone: may be either ductile or brittle	• General or localized surface attack (pitting, cracking) • Selective phase attack • Thickness and morphology of corrosion scales	• May show localized distortion at surface consistent with direction of motion • Identify embedded particles	• Microstructural change typical of overheating • Multiple intergranular cracks • Voids formed on grain boundaries or wedge-shaped cracks at grain triple points • Reaction scales or internal precipitation • Some cold flow in last stages of failure
Contributing factors	• Load exceeded the strength of the part • Check for proper alloy and processing by hardness check or destructive testing, chemical analysis • Loading direction may show failure was secondary • Short-term, high-temperature, high-stress rupture has ductile appearance (see creep)	• Load exceeded the dynamic strength of the part • Check for proper alloy and processing as well as proper toughness, grain size • Loading direction may show failure was secondary or impact induced • Low temperatures	• Cyclic stress exceeded the endurance limit of the material • Check for proper strength, surface finish, assembly, and operation • Prior damage by mechanical or corrosion modes may have initiated cracking • Alignment, vibration, balance • High cycle low stress: large fatigue zone; low cycle high stress: small fatigue zone	• Attack morphology and alloy type must be evaluated • Severity of exposure conditions may be excessive; check: pH, temperature, flow rate, dissolved oxidants, electrical current, metal coupling, aggressive agents • Check bulk composition and contaminants	• For gouging or abrasive wear: check source of abrasives • Evaluate effectiveness of lubricants • Seals or filters may have failed • Fretting induced by slight looseness in clamped joints subject to vibration • Bearing or materials engineering design may reduce or eliminate problem • Water contamination • High velocities or uneven flow distribution, cavitation	• Mild overheating and/or mild overstressing at elevated temperature • Unstable microstructures and small grain size increase creep rates • Ruptures occur after long exposure times • Verify proper alloy

(a) Failure at the time of load application without prior weakening. (b) Failure after a period of time where the strength has degraded due to the formation of cracks, internal defects, or wastage. (c) EDS, energy-dispersive spectroscopy

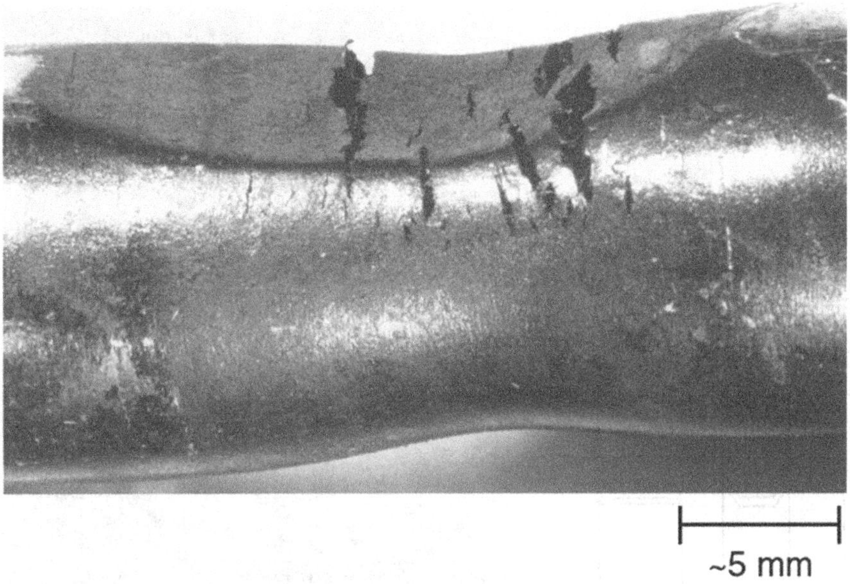

Fig. 12.1 Example of tensile fracture with stretching in fracture region in a bolt with cracks that opened up during bolt tensioning. Source: Ref 1

Tensile fractures are failures in which the tensile load exceeded the tensile strength of the material in some region of the part. In this situation, the material may exhibit some stretching or necking, depending on the ductility of the material or the geometric configuration of the parts and stress distributions. If the tensile fracture is ductile, it typically exhibits some necking or stretching in the region of the fracture site (Fig. 12.1). This indicates some material stretching and thinning prior to fracture.

Brittle fractures are failures in which the material fails in a brittle manner. Brittle fractures typically manifest themselves as a sudden fracture with little or no deformation prior to cracking. Structural brittle fracture can be catastrophic, with significant energy released during fracture (frequently with a loud crack).

Brittle fractures exhibit sharp or flat fracture surfaces, similar to what one may observe in a broken piece of glass. In terms of macroscopic profile of the fracture, a fracture is brittle if it is on a plane normal to the maximum normal stress (condition 4 in Fig. 12.2). A fracture is considered to be macroscopically ductile when the fracture surfaces are inclined to an imposed load (slant fracture or plane stress). Toughness of the materials is higher under conditions of ductile fracture, and some additional work may be expended in work-hardening deformation under load. A fracture surface displaying both types of planes can be described as a mixed-mode fracture or, alternatively, by indicating the presence of shear lips on the fracture surface.

Causes of brittle fracture may be geometric, such as large preexisting crack sizes or section constraints (roughly outlined in Fig. 12.2). Material

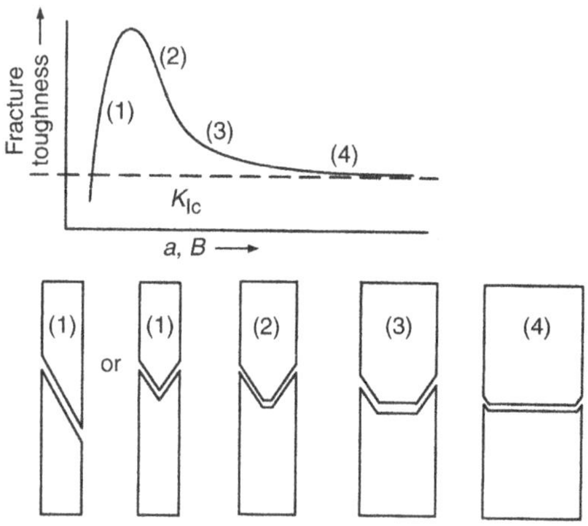

Fig. 12.2 Schematic of fracture surfaces for an inherently ductile material with variation in fracture toughness and with variations in section thickness (*B*) or preexisting crack length (*a*). With an increase in either section thickness (*B*) or preexisting crack length (*a*), conditions of plane strain (brittlelike fracture) develop first along the centerline and result in a flat fracture surface. With further increases in section thickness or crack size, the flat region spreads to the outside of the specimen, decreasing the widths of the shear lips. When the minimum value of plane-strain toughness (K_{IC}) is reached, the shear lips have very small width. Source: Ref 2

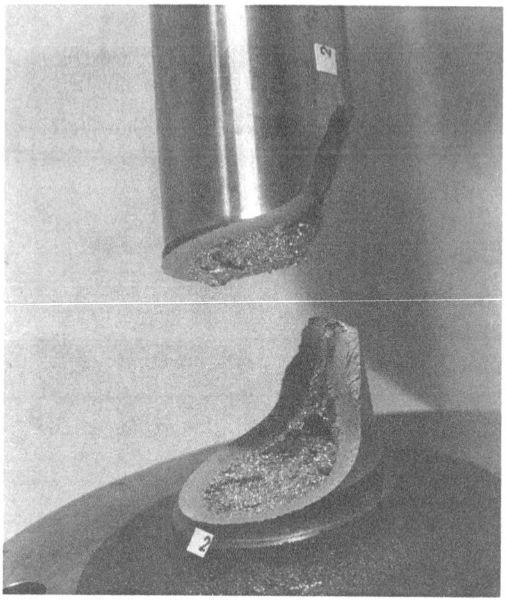

Fig. 12.3 Brittle fracture of a modified SAE 1050 (0.50% C, 0.95% Mn, 0.25% Si, 0.01% S, and 0.01% P) axle shaft due to single-bending impact load in a lab test. The hot rolled and upset shaft had an induction-hardened case (60 HRC) with a softer core (20 HRC). Failure occurred at the flange end.

factors can also cause brittle fracture. If a material is too hard, it may be susceptible to failing in a brittle manner (Fig. 12.3). The fracture surface in Fig. 12.3 is not flat, because the load was from a bending impact. However, the fracture in the hardened outer case is shiny (indicative of brittle fracture, Table 12.1), while the fracture surface in the softer core has a fibrous surface, indicating deformation during fracture.

Brittle fracture may be attributed to a variety of possible root causes, including incorrect design, work hardening, and embrittlement of otherwise ductile materials due to improper processing or exposure to certain liquids or gases. For example, hydrogen embrittlement of high-strength steel can occur during welding or heat treating as a result of hydrogen atoms migrating into steel. The fracture from hydrogen embrittlement may occur hours after the load is applied.

The failure analysis team should assess if the design adequately controls hardness. (The specified hardness may be too high, or the design may not specify hardness.) Materials analysis can determine if the failed item was constructed using the specified material, if it was correctly heat treated, or if embrittlement occurred for some other reason. The corrective actions here may involve a design change to lower the component hardness if the hardness is inappropriately specified, modifications to the process, or the use of a different material.

Compressive failures occur when a component is loaded in compression and crushes or buckles. These failures may exhibit bowing, stretching, and fractures of the surfaces parallel to the load path. Compressive failures can result from loads exceeding the design limits, from the material being too soft, or from the part not meeting its drawing requirements. Materials analysis can determine if the component was fabricated using the correct material. Hardness testing can determine if the hardness is correct. Dimensional inspection can determine if the component met its drawing requirements. Stress analysis, and perhaps load testing, can determine if the design is adequate. Corrective actions may include design changes (either changing the dimensions or the material) or process changes (if the failed component did not meet drawing requirements).

Fatigue failures can occur if a part is subjected to repetitive stress cycles that ultimately cause it to fail in fatigue. A common example (one that most people can readily understand) is simply bending a paper clip back and forth until it breaks. The back-and-forth bending is called a stress reversal, but a part need not be subjected to such obvious stress reversals to experience fatigue failures. Rotating shafts, for example, always experience stress reversals due to dimensional variation. Vibration can induce stress reversals. Components subjected to these types of stress reversal may experience fatigue failures if they are not designed to withstand the reversals they experience in normal use, if the stress reversals are higher than anticipated, or if the stress reversals continue for periods longer than anticipated.

Fatigue fracture surfaces often, but not always, exhibit characteristic "beach" marks (Fig. 12.4), so called because the marks are similar to those left by waves as they crash on the shoreline of a beach. Beach marks represent an initial fracture front, followed by a progressively advancing fracture front as succeeding stress reversals occur. Design analysis determines if the design is adequate, and dimensional and materials analysis determines if the component met the drawing requirements. The failure analysis team should determine if the stress cycles experienced by the failed part were within the predicted fatigue life for the part.

Deformation

Deformation refers to the occurrence of permanent plastic strain when stresses exceed the elastic limit of the material. Every mechanical part subjected to a load experiences some deformation. Most parts are designed such that the load and resultant deformation are well below the elastic limit of the material. That is, when stresses are below the elastic limit, the

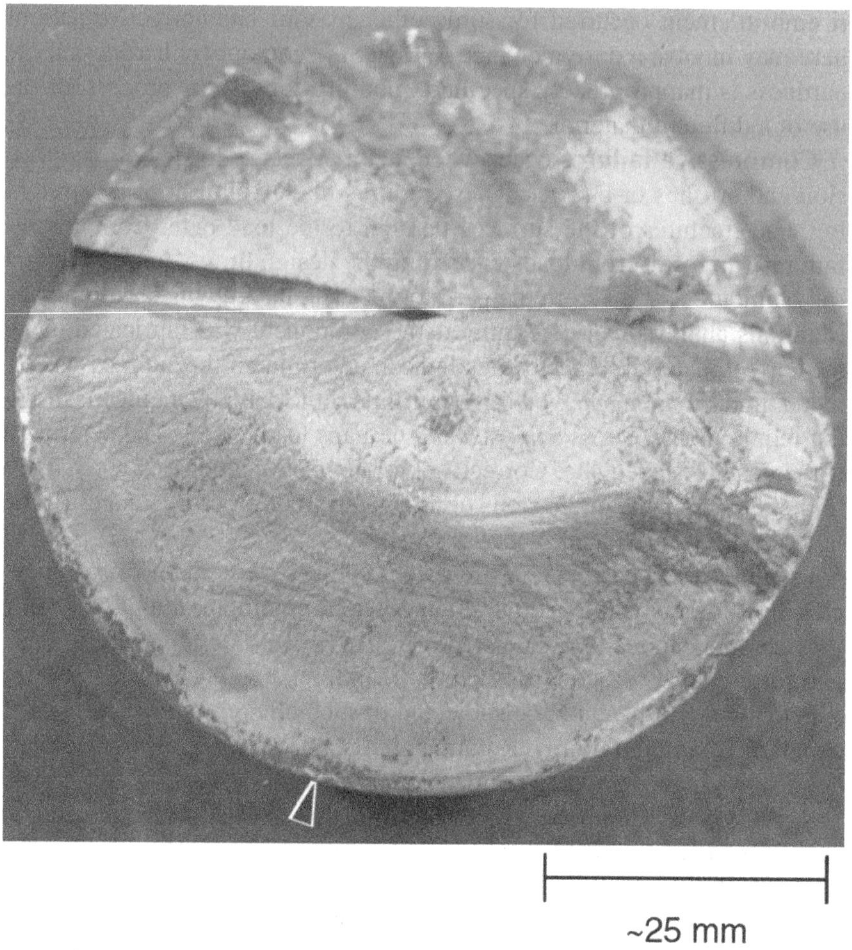

Fig. 12.4 Fatigue fracture surface appearance of a failed crankshaft, showing beach marks on the lower part. The origin of the primary fracture is indicated by the arrow. Source: Ref 1

part returns to its undeflected state when any deflection or deformation under load is removed.

If the amount of bending is so small as to not interfere with the system function, no failures will result. In some cases, however, the fault-tree analysis may identify situations in which component deflection may inhibit system function or induce a system failure. What complicates this analysis is that if the parts have not been deflected beyond their elastic limit, they will return to their undeflected configuration when the load has been removed.

If the fault tree hypothesizes interference induced by deformation as a failure cause, the failure analysis team can examine suspect parts for witness marks that would indicate the parts interfered with each other. Another approach is to photograph the parts under varying load conditions, including hypothesized worst-case load conditions, to determine if sufficient deformation bending occurred. Sometimes, multiple photographs or motion pictures can pick up deformation that may not otherwise be visible. Finally, the parts should be dimensionally inspected. A dimensional inspection can detect if the deformation was severe enough to deform the parts.

Mechanical Interference and Wear

The fault tree may identify conditions in which mechanical interference can prevent proper system operation, induce inadvertent operation, or accelerate wear. To address this concern, the failure analysis team should dimensionally inspect suspect parts to confirm engineering drawing compliance. Under certain tolerance conditions, it may be possible for conforming parts to interfere, to have excess clearance, or to have inadequate engagement. The failure analysis team can visually and microscopically inspect parts for interference. If interference occurred, it may leave witness marks. Figure 12.5 shows this condition.

The failure analysis team should ask for a tolerance analysis to assess nominal dimensions and tolerances to evaluate potential component interference, inadequate engagement, or such related hypothesized conditions. The tolerance analysts should consider worst-case conditions within the components specified tolerances to determine if dimensional stackups could induce the undesired condition. In many cases, engineering designers do not rigorously evaluate all tolerance conditions under worst-case conditions. Stated differently, parts may be completely conforming to drawing nominal dimension and tolerance requirements yet still interfere or not

Fig. 12.5 Mechanical interference witness marks. This component seized as a result of mechanical interference with other components. Witness marks show where the interference occurred.

mate properly. (In such a situation, the design is defective.) The failure analysis team should identify specific suspected conditions in order to appropriately focus the tolerance analysts' efforts.

Wear can induce failures. It can result from the types of interference described previously or from inadequate lubrication. Wear indications include abrasions or other wear patterns on mating parts, the presence of material abraded from mating parts in the lubricant, or the buildup of material abraded from mating parts at the part interface. The presence of wear indicators is usually visible under magnification. The presence of abraded material in lubricants can often be determined by a spectroscopic analysis of the lubricant. Spectrographic lubrication analysis may also reveal lubricant anomalies (e.g., use of the wrong lubricant) or the presence of contaminants indicating failures in other areas (e.g., the presence of water, materials from other mating components, etc.). Sometimes, a simple visual examination of the lubricant shows the presence of abraded materials.*

Fretting is a special type of mechanical interference that typically refers to vibration-induced wear. Corrosion may also be present in such cases, because the wear exposes fresh material that subsequently oxidizes. Residual powders may exist where fretting has occurred. (A red powder may appear with ferrous materials, or a black powder may appear for aluminum materials.)

Mechanical Disconnects

Many designs incorporate threaded fasteners for attaching components. These include nuts, bolts, screws, and other threaded joints. Threaded fasteners can disconnect for any of several reasons, including vibration, inadequate torque (during assembly or maintenance), missing threaded fastener security systems (discussed subsequently), and, in some cases, overtightening.

In many cases, a failed system can be inspected to determine if the suspect threaded fastener is loose or disconnected. In cases in which the failed system is no longer available, it may be possible to inspect similar available systems for the hypothesized aberrant condition.

Threaded fasteners should be tightened to a torque value. If no torque has been specified, the torque to which a fastener should be tightened is inadequately controlled by the product design or process, and corrective action should include specifying the correct torque.

If a torque value is specified, determining if it has been met can be challenging. The failure analysis team can check the torque required to loosen the fastener as an indication of the torque to which it has been tightened and compare this to published values for the size of the fastener. The fail-

*The failure analysis team should realize that the presence of abraded materials in a lubricant is often an expected, normal condition. In such cases, an abnormally high abraded-material content indicates excessive wear.

ure analysis team can also note the radial position of the fastener head, loosen it, and then retorque it to the correct torque. If the fastener head returns to the same position, it may indicate that the fastener was properly installed (but this is not conclusive).

If a threaded fastener is overtorqued, it may exhibit deformation. Overtightening can also induce threaded fastener failures. The deformation and overstressed condition can result in the fastener head popping off or the fastener shaft fracturing. The overtightened condition can result in thread deformation, which subsequently decreases the grip on the fastener and allows it to back out.

Another way to determine if threaded fasteners are being properly tightened is to interview and observe the technicians who install the fasteners. Based on this information, the failure analysis team can assess the likelihood that the specified torque values are being met. (This is also not conclusive, but it is a good indication.) If the technicians are not using torque wrenches (but the drawings specify torque values), the fasteners are almost certainly not in compliance with their torque requirements. If the torque wrenches are not calibrated in accordance with a calibration schedule, it is likely that the installation torque is nonconforming.

Threaded fasteners frequently include secondary security features to lock them in place. These may include adhesives to secure the fastener (e.g., LocTite), lock washers, and safety wire. The failure analysis team should confirm that these secondary features are being used if they are required by the engineering drawings and that the right parts or materials are being used. There are various types of LocTite, for example, and the failure analysis team should confirm that the type specified by the engineering drawing is the type being used.

Composite Material Failures

Both metallic and composite materials are widely used in a variety of applications, and these are frequently joined with epoxy or other adhesives. The term *composite* may refer to metallic items bonded together with adhesives, or it may refer to resin-based materials. Either type of material can experience disbonds. In addition to disbonds, resin-based materials can also experience crazing, cracking, structural failures, and delaminations.

Composite materials are usually held together with adhesive bonds. If disbonds occur, several conditions should be assessed. Bondline dimensions (bond thickness, length, and width) are critical. This situation is complicated by the fact that engineering drawings for bonded assemblies often do not specify bond dimensions. (Instead, the drawings specify the dimensions of the bonded parts.) The resulting bond dimensions are based on the dimensions of the interfacing parts. If the bond thickness, length, or width is too small or too large, a disbond can result. In this situation, the failure analysis team should request a tolerance analysis to show minimum bond dimensions under worst-case conditions. The nominal bond dimensions may be adequate, but at tolerance extremes, the bond dimensions can create a situation in which a failure may occur.

In some cases, adhesives require a specified temperature for a specified time for curing. If the cure specifications are not met, a weak bond results. The failure analysis team should assess how curing is controlled. They may find that the specified cure conditions were not met or that the manufacturing instructions are too vague with respect to cure time. Corrective actions should include appropriate cure-time specifications and monitoring.

If all dimensions are in conformance and the bond joint fails, the bond design may be inadequate. The failure analysis team should consider checking with the epoxy manufacturer to identify recommended bond dimensions for anticipated loads and if the specified adhesive is appropriate for the application.

Cleanliness is essential for a good bond. If any contaminants are present, the bond joint will degrade. This is usually apparent through an examination of the failed bond joint. If the adhesive or epoxy appears to stick almost entirely to one side of the bonded joint and the mating part is relatively free of epoxy in the area where the joint failed, it is highly likely that the mating part surface was contaminated (Fig. 12.6). This situation is referred to as an adhesive failure. The failure analysis team should examine the working conditions in which the bond joint was applied. If the workstation has lubricants or silicone in the area, it is an indication that the bond area was contaminated. The failure analysis team can also use Fourier transform infrared (FTIR) spectroscopy (as outlined in Chapter 11) to assess if contaminants were present. FTIR is commonly used for this purpose, and it is particularly well suited for confirming silicone presence.

If the adhesive or epoxy bond material is present on both of the bond mating surfaces, this indicates a failure in the adhesive or epoxy material. (This is called a cohesive failure.) Such a condition does not represent an adhesion problem induced by a contaminated surface, as described earlier. Instead, the bonding agent was too weak for the loads it experienced. This failure can be induced by several causes:

- The loads may have exceeded the strength of the adhesive (Fig. 12.7).
- The adhesive or epoxy may be too old (i.e., it may have exceeded its shelf life).
- If epoxy was used as the bonding agent, it may not have been properly mixed, or it may be an incorrect epoxy for the application.
- If epoxy was used, it may have exceeded its pot life (its useful application time after it was mixed or removed from the refrigerator*).

---

*Most epoxies are either mixtures of materials that are combined and mixed just prior to use or premixed epoxies that must be kept refrigerated until ready for use. The epoxy bond joint will be weak if the mix ratio is incorrect, the mix is not completely mixed, the refrigeration temperature is too high, or the premixed epoxy is kept out of the refrigerator for too long before use.

Fig. 12.6 Adhesion failure. The large area on the upper right side of this bonded material shows no adhesive. All of the adhesive adhered to the left side of the bonded material.

- If epoxy was used as the bonding agent, voids or bubbles in the epoxy indicate that it was mixed too aggressively prior to application.

The failure analysis team should check for all of the aforementioned indications, confirm that the technicians assembling the system used the correct adhesive or epoxy, and ensure that the adhesive or epoxy specified is appropriate for the application by contacting the epoxy manufacturer.

Composite materials are frequently provided as "prepreg." Prepreg materials are woven fibers preimpregnated with uncured resin. Such materials must be kept refrigerated prior to use. After assembly in a mold or other composite material tool, they are cured for a specified time at a specified temperature. Structural failures in such materials can occur for several reasons:

- The design may be inadequate (i.e., the loads exceed the assembly strength). Stress analysis, testing, or both can be used to determine if this condition exists.

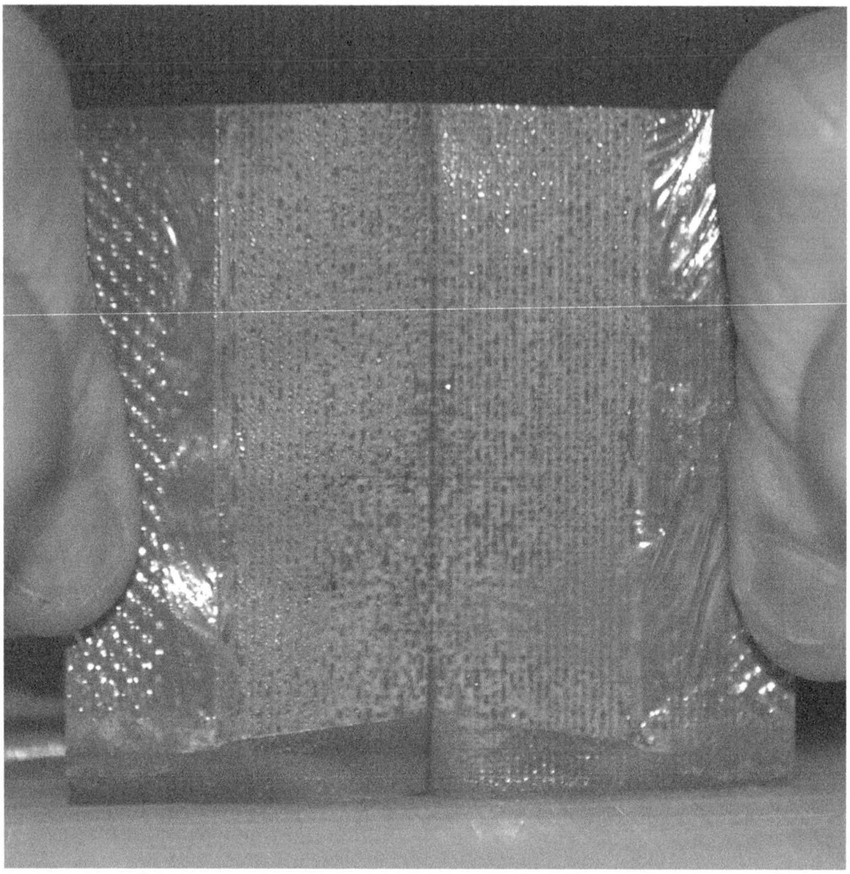

Fig. 12.7 Cohesive failure. In this specimen, the adhesive adhered to both sides of the bonded material; it failed in the adhesive layer.

- The cure may not have been appropriate. The failure analysis team can use the same approach as outlined earlier for adhesives requiring a cure to determine if the cure specification is adequate and if it was met.
- The prepreg material may not have been kept appropriately refrigerated. The failure analysis team should investigate the material storage history and the organization's storage practices.
- The prepreg material may have an expired shelf life or pot life. The failure analysis team should similarly investigate the organization's operational practices.
- Contaminants may have degraded the material. As was the case for adhesives, the failure analysis team should investigate material and tooling work and storage areas for the presence of contaminants.

Composite material crazing, cracking, and other similar failures are addressed in the following section.

Plastic Component Failures

Plastic component failures can be analyzed using many of the same techniques as metal failures, including visual and microscopic examination, surface fractography, and materials analysis using the technologies described in the preceding chapters. Plastic components are also susceptible to unique failure modes, as explained subsequently.

Aging can induce failures in plastic materials. Age-related failures can be the result of exposure to an environment that accelerates the plastics aging process (chemicals that interact with the material, moisture, heat, or combinations thereof), repeated stress cycling, prolonged exposure to a steady load, or prolonged exposure to sunlight or ultraviolet radiation. As a result of exposure to such environments, plastic materials may become more pliable and fail in a ductile manner, or they may become less plastic and fail in a brittle manner. The plastic material supplier may be able to help in determining if the material is compatible with its intended environments.

Deformation in plastic parts can result from the load exceeding the part strength. Temperature and load are both critical factors here. Load is critical for obvious reasons; temperature is critical because plastic materials are susceptible to significant decreases in strength as temperature increases, as well as at low temperatures. For a thermoplastic material, strength decreases rapidly as the melting temperature is approached. The same phenomenon occurs in thermoset plastics as the glass transition temperature is approached. At low temperatures, many plastics are more susceptible to brittle failures. Some plastics have a tendency to creep (change dimensions) under steady load. When plastic components deform or fracture in a brittle manner, the failure analysis team should consider the temperatures the component experienced both in operation and in storage. It usually makes sense to contact the plastic material source to confirm that the material is compatible with the operating and storage environments.

Crazing is a failure mode in which microcracks and sometimes color shifts develop in a plastic material. Stress, a hostile environment (exposure to chemicals that degrade the material), or a combination of stress and a hostile environment can induce crazing. Aging can also induce crazing. The presence of microcracks and color shifts may be undesirable for aesthetic reasons, or the microcracks can progress to larger cracks that ultimately lead to part failure. Again, the plastic material supplier can sometimes assist in evaluating the susceptibility of a material to crazing.

In some cases, the material supplier may not be able to assist the failure analysis team in assessing a plastic component failure. In such cases, commercial laboratories (as described in Chapter 11) may be helpful. In other cases, the failure analysis team should consider an experiment using the techniques discussed elsewhere in this book to evaluate suspected environmental effects.

Ceramic Component Failures

Ceramics and high-performance ceramics are used in many designs. Conventional ceramic components have been in use for many years. High-performance ceramics are more recent, and they offer a combination of low weight, high strength, and low radar signature. (For example, the new F-35 fighter aircraft uses ceramic turbine blades in its engines.) Such engineered materials can also fail by fracturing, leaking, or failing to meet their requirements.

When ceramic parts fracture, they typically do so in a brittle manner due to the nature of ceramic material. Sometimes, if pieces of the failed component are available, the part can be reassembled to assess how the fracture progressed. This technique may be useful for determining the loads experienced by the part before it failed as well as the fracture point of origin and direction. Such information may be useful in determining if the fracture originated from an impact load or repeated stress cycling, if it had machining discontinuities or other stress raisers, if it met drawing requirements, or if it experienced localized or uneven heating (which can exacerbate residual stresses or create stresses that exceed the material strength). In many cases, metal parts that interface with ceramic parts can induce localized regions of high stress, leading to failure (as may occur, for example, where a threaded fastener is used to secure a ceramic component).

In some cases, ceramic components may require specialized failure analysis techniques and the assistance of a commercial laboratory, or the ceramic component manufacturer may be helpful. In other cases, many of the techniques discussed in this and preceding chapters can be used to evaluate materials, assess operating environments, examine fractures, wear patterns, and witness marks, and so on.

Electrical and Electronic Failures

In electromechanical or electrical systems, the fault-tree analysis frequently identifies electronic component or subassembly failures as potential causes of a system failure. When this occurs, several techniques are useful for determining if the hypothesized failure cause occurred.

Electronic Subassemblies. Fault-tree analyses often show subassembly failures as undeveloped events. (The subassembly is treated as a "black box.") These failures are relatively easy to confirm. The failure analysis team can define the required inputs and outputs of the electronic subassembly. It is then a simple manner to provide the required inputs and measure the outputs to determine if the subassembly is performing as it is supposed to or if it has failed.

In some cases, the failure analysis team can rely on the device appearance and its smell. Many electrical and electronic component failures are associated with overheating and burning. If the device shows traces of soot or smells as if something in it has burned, it has probably failed. If the subassembly has a cooling fan, it should run without making any unusual noises. If the subassembly has cooling vents, they should not be clogged with dust or other foreign material. The presence of any of these conditions indicates the device has failed.

Circuit Card Assemblies. In many cases, the fault-tree analysis shows circuit card assembly failures as potential failure causes. Circuit cards can have many failure causes, including:

- Individual component failures
- Component connection failures
- Delaminations
- Mechanical fractures of components or electrical leads
- Use of the wrong component
- Reversed component installation
- Overheating
- Bent pins
- Short circuits

When examining circuit card assemblies for the presence of any of the aforementioned conditions, the failure analysis team can use the following criteria:

- All electrical and electronic components should be present; they should be installed in the correct orientation; they should be of the correct type; and they should be connected.
- The circuit card assembly should not have any evidence of burning. There should be no burn marks or burned smell.
- The circuit card assembly should not have any evidence of mechanical damage, such as scratches, bent pins, delaminations between the circuit card layers, or bending.
- When connected and energized, the circuit card assembly should not make any unusual noises, such as hissing or crackling.
- All fuses on the circuit card assembly should be intact.
- All solder joints on the circuit card assembly should be intact, and the solder connections should be sound. There should be no solder spatter around the soldered connections.
- The circuit card assembly should be free of corrosion.
- The circuit card assembly part number and revision level should be correct.

In some cases, the failure analysis team can use the aforementioned inspection criteria to isolate a circuit card assembly defect to a specific electrical component on the card. In other cases, it may be possible to isolate the failure to a specific component by lightly touching each component when the board is energized to ascertain if any are unusually hot. (However, caution is required to prevent electrical shock or electrostatic discharge damage.)

As was the case with electrical or electronic subassemblies (as explained previously), the failure analysis team can check circuit card assembly performance by providing required inputs and measuring the circuit card output. The failure analysis team can also replace the circuit card with a known good one to determine if the suspect card is functioning properly.

Electrical Component Failures. If the fault tree hypothesizes a failure cause to a specific electrical or electronic component, perhaps the easiest way to confirm if the component is defective is to replace it with a known good one. If the known good component works, the failure may be due to the replaced, presumably defective component.*

In addition to this approach, the failure analysis team can inspect components to determine if they are defective using techniques described subsequently. Sometimes, the defective component is visible (Fig. 11.1). In many cases, these inspections also offer insight into the component failure cause. (For example, if solder connections are spattered, the component may have been subjected to excess energy, as subsequently explained in more detail.) The failure analysis team can also check the component with a multimeter to determine if it has failed open or has a short circuit.

Integrated circuit (IC) components may show evidence of overheating if an overvoltage condition occurred. If an overvoltage condition existed, the component shows melted silica, melted or spattered solder connections, or melting of other IC materials. The IC may have holes in it and may show melting, vaporization, and redeposits of molten IC materials as they cooled in the area surrounding the melting. Examination under magnification is typically required to make these determinations. A burned smell to the component is another indication of overheating.

Resistors typically fail due to high temperature as a result of overcurrent or overvoltage conditions. Evidence of overheating includes the same smell test described previously. The resistor may show discoloration, burn marks, blistering, or disintegration. The resistor may have become hot enough to melt its connections, so the failure analysis team should look for melted leads. Sometimes, no marks are present, or the resistor may still work (but at a reduced efficiency). The failure analysis team can test the resistor to confirm that its resistance value is within tolerance limits.

Resistors can also fail as a result of mechanical overstress conditions. A cracked resistor body, bent pins, or changes in the resistance value indicate that an overstress condition occurred.

Capacitor failure modes include leaking (an electrical loss of stored energy or actual leaking of the capacitive medium), reduced capacitance, or

*The failure analysis team must exercise caution when doing this. If the failure is intermittent or due to statistical variations in load and strength, the failure may not recur even if the component had not been replaced. The concept of randomness is addressed in more detail in Chapters 16 and 17.

Fig. 12.8 Defective solder joints. Note the poor solder joints on the left side of the component. Such defects can induce open circuits or create a short circuit if the solder forms a bridge between components. Solder spatter of the type shown here can also be induced by excess energy and overheating.

short-circuiting. Figure 11.1 shows a capacitor that ruptured and leaked, which was apparent without magnification. Capacitor failures can be induced by overheating, subjecting the capacitor to excess voltage, exposing the capacitor to excessive vibration or temperature cycling (which can induce cracks and reduce capacitance), or aging. As described earlier, the easiest way to determine if a capacitor is defective is to replace a suspect capacitor with a known good one. In some cases, the failure analysis team may wish to couple the capacitor with a resistor of known value and then measure the voltage rise over time to determine if the capacitor is operating correctly.

For all hypothesized electrical and electronic component failures, the failure analysis team should check to make sure the correct component was installed, the component was installed in the correct orientation, and the contacts are intact. The team should also check for any evidence of mechanical damage. Finally, the team should determine the component age, because older components are more likely to fail.

Connection Failures. Frequently, electrical or electronic component failures are not failures of the component but are connection failures instead. This can occur in the case of components soldered to a circuit card if the wave or manual soldering is defective (Fig. 12.8). The solder connection may have originally been sound, but excess energy through the connection can melt the solder, inducing a disconnected condition. Connection failures can also result in a short circuit, either by solder bridging across connection points or by bent connector pins contacting each other.

In many other cases, connection failures are due to mechanical fractures that may occur as the result of tensile, bending, or compressive loads, as described earlier for mechanical failures. In such cases, microscopic examination of the failed leads can reveal the mode of fracture, and from that, the failure analysis team can deduce the failure cause.

Failure Analysis with No Hardware. In some cases, the failed system may not be available for the failure analysis team to examine. This occurs, for example, if an aircraft is lost, a bomb detonates prematurely, a missile misses its target, or the product fails in a remote location and cannot immediately be returned for examination.

This situation is not as difficult as it may appear. Recall that the first steps in a system failure analysis are to gather as much information as possible, define the problem, prepare a fault-tree analysis and an FMA&A, and then assign actions to failure analysis team members to rule in or rule out hypothesized failure causes. Previous discussions pointed out the advantages of performing these first few steps prior to examining any equipment. Doing this ensures that the failure analysis team knows what information to seek. If the failure analysis team begins to examine failed hardware before doing the aforementioned, they may overlook or destroy critical evidence. Many activities associated with a systems failure analysis do not involve examining the failed hardware. These other activities include pedigree, change, contaminant, environmental, process, and design analyses as well as the design of experiments (discussed in Chapter 17). These analyses can reveal much about the likelihood of hypothesized failure modes without examining the failed system.

If the failed system is not available, it does not mean the failure analysis team cannot examine representative hardware. The failure analysis team can examine inventory (both in-house at suppliers), work in process, and other fielded systems. Frequently, many of the hypothesized failure modes are also present in these items.

Summary

Appendix A summarizes component failure modes and how to confirm or refute their existence based on the material presented in this and the preceding chapter.

The failure analysis team should not inspect components or subassemblies from a failed system until after the team has identified potential failure causes, prepared the FMA&A, and defined specific areas to examine. Launching into an inspection of a failed system without first defining what to look for may destroy valuable evidence or reduce the knowledge to be gained from the failed hardware.

In some cases, the failure analysis team may find that all components and subassemblies meet all drawing and specification requirements. In these cases, the design is inadequate.

Mechanical components are comparatively easy to evaluate. Surface fractography usually shows if failures occurred as the result of the material being too brittle, if the tensile or compressive loads exceeded the material capability, or if fatigue induced the failure. The failure analysis team can use a variety of approaches to determine the causes of fractures, bending, interference, or mechanical disconnects.

Contamination, incorrect mixing, or poor process control can induce bond-joint failures. Exceeding the bonding agent strength, shelf life, or pot life can also induce bond-joint failures. The failure analysis team can determine the cause of these types of failures with separated surface examinations, materials analysis tools, process assessments, and environmental evaluations.

Electrical and electronic subassemblies and components can be evaluated by replacing the suspect item with a known good one. The failure analysis team can use smell and visual examinations to determine if components were subjected to mechanical overstresses or excess heat. In some cases, visual or microscopic examinations can determine if overheating was induced by overvoltage or overcurrent conditions.

In some cases, hardware from the failed system may not be available. This is not as big a handicap as it may appear. Many of the activities associated with a systems failure analysis do not require examination of the failed hardware, and other hardware (fielded systems, work in process, and inventory) can be examined to confirm the existence of hypothesized failure causes.

A word of caution is in order here. In all cases, when component failures are confirmed by replacing the suspected components with known good ones, the failure analysis team should attempt to identify why the failed component failed and eliminate the failure cause. Simply replacing the component is not an appropriate corrective action. If this failed component failure cause is not identified, the new component is likely to fail for the same reasons as the original component.

Example for Group Discussion. Many systems failures are induced by mechanical or electronic failure causes in components or subassemblies provided by suppliers. Is simply replacing the failed component with a known good one sufficient, or should the failure analysis team take additional steps?

If a failed item provided by a supplier met all requirements on the purchase order, the specifications, and the engineering drawings, what does this imply? Who bears responsibility in this case?

REFERENCES

1. I. Le May, Examination of Damage and Material Evaluation, *Failure Analysis and Prevention,* Vol 11, *ASM Handbook,* ASM International, 2002, p 352

2. W.T. Becker, Fracture Appearance and Mechanisms of Deformation, *Failure Analysis and Prevention,* Vol 11, *ASM Handbook,* ASM International, 2002, p 566

SELECTED REFERENCES

- *Dissimilar Metals,* MIL STD 889, U.S. Department of Defense, 1976
- D. Galler and R.A. Blanchard, Failure Analysis of Components, *Electronic Failure Analysis Handbook,* P.L. Martin, Ed., McGraw-Hill, New York, NY, 1999
- J.I. Goldstein and D.E. Newbury, *Scanning Electron Microscopy and X-Ray Microanalysis,* 2nd ed., Plenum Press, New York, NY, 1992
- C.J. Hellier, Radiographic Testing, *Handbook of Non-Destructive Evaluation,* C.J. Hellier, Ed., McGraw-Hill, New York, NY, 2001
- P.L. Martin, Semiconductors, *Electronic Failure Analysis Handbook,* P.L. Martin, Ed., McGraw-Hill, New York, NY, 1999
- D.J. Wulpi, *Understanding How Components Fail,* ASM International, 2000

CHAPTER **13**

Leaks

BASED ON EXPERIENCE with many systems, a surprisingly large number of system failures involve leaks. Leaks can cause any of several system failures:

- Leaks can permit water intrusion into critical areas, with the water intrusion resulting in corrosion, electrical short-circuiting, component damage, or hydrostatic lockup of moving parts.
- Leaks can allow inert gases to escape from chambers dependent on the inert gas to prevent corrosion or other chemical reactions.
- Leaks can allow the escape of lubricating fluids, which can result in overheating, friction-induced damage to mating parts, lockup of moving parts, and damage resulting from the escaping lubricant.
- Leaks can allow hazardous liquids or gases to escape.
- Leaks can result in loss of pressure or creation of excess pressure in areas where the liquid or gas migrates.
- Leaks can induce failure to function or inadvertent function in hydraulic or pneumatic logic mechanisms.
- Leaks can result in a loss of pressure in backup systems.
- Leaks can result in gas or liquid mixing where none is allowed.
- Leaks can create stress raisers in bonded joints or welded joints and provide a crack-initiation site.
- Leaks can reduce system efficiencies if the leak path is around a filter or other control feature.
- Leaks can result in test failures in tests where freedom from leakage is an acceptance characteristic.

Leaks can occur as the result of several failure causes. These are reviewed in detail as follows.

Gasket Leaks

A gasket is made of a compressible material squeezed between two mating parts that are usually held together by threaded or other fasteners. The concept is that the gasket is squeezed evenly and tightly enough between the two mating surfaces to deform the gasket and form a seal between the mating surfaces. Causes of gasket leaks include:

- Anything that prevents the gasket from deforming enough to form a seal with each mating surface can induce a leak. One such cause is any imperfection on either mating surface, such as a scratch, gouge, or surface finish that is too rough. Contaminants caught between the gasket and the mating surfaces can allow a leak path. Sometimes, gaskets are used with sealants, and the sealant flows into minute gaps between the gasket and its mating surfaces. If the sealant is not used, sealing may not occur. If the mating surfaces previously used sealant and the parts are disassembled, the mating surfaces may not have been adequately cleaned. If the parts are not adequately cleaned, old, hardened sealant may remain and act as a leak-inducing contaminant. If the old sealant is cleaned from the mating surfaces too aggressively, the surface may be gouged or scratched during cleaning, providing a leak path.
- If the gasket material is torn or otherwise damaged, a leak path may result. This could be present in a new gasket, or someone may attempt to reuse an old gasket.
- The gasket may have the wrong dimensional configuration. This can occur if the gasket is too thick or too thin or if it does not have the right outline for the surfaces to be sealed. The failure analysis team should be alert to the possibility suggested by gasket drawings that have been revised. If the wrong gasket revision level is used, an earlier revision may have different dimensions.
- If the gasket has been fabricated from the wrong material, the fluid being sealed may damage the gasket, resulting in a leak. This can occur from chemical incompatibility or from temperature.* When evaluating a gasket leak, the failure analysis team should consider a gasket material analysis to confirm the material. The failure analysis team should also seek independent input to confirm that the design specifies a material appropriate for the environment.
- Corroded mating surfaces may create a leak path. The failure analysis team should inspect the mating surfaces for corrosion and confirm that the interfacing materials are not chemically reactive. Corrosion-inducing reactions may occur as the result of incompatibilities between the gasket, the mating surfaces, the fluid being sealed, and the threaded fasteners holding the mating surfaces together. The corrosion may be chemical or galvanic.

*The reader may review the aircraft accident described in Chapter 9, which outlined a situation in which an aircraft caught fire and crashed into the ocean. The cause was determined to be an oil filter gasket that melted, allowing oil to spill out over the engine exhaust manifold. A fire resulted, the aircraft crashed, and a woman died. The gasket had been fabricated using an incorrect material. (The material was not rated for the temperatures experienced on the aircraft engine.)

- Inadequate or excessive gasket squeeze by the fasteners holding the mating surfaces can induce leakage. If the threaded fasteners are torqued too lightly, inadequate gasket deformation may occur, inducing leakage. If the mating surfaces holding the gasket together are squeezed too tightly, a leak may not occur immediately but may appear later if the gasket material splits or deforms over time. If the mating surface threaded fasteners are not tightened in a load-distributing sequence, asymmetric gasket squeeze may occur, resulting in leakage.
- In some cases, attempts at correcting gasket leaks with chemical additives can do more harm than good. These chemical additives typically are designed to induce gasket swelling (or expansion) in an attempt to force a seal. In some cases, however, the swelling creates a leak path, and the additive aggravates rather than corrects the leak.

O-Ring Leaks

O-rings are different than gaskets in that they are captured in an O-ring channel formed by the two mating surfaces to be sealed. The two mating surfaces contact each other directly, so they cannot exert excessive squeeze on the O-ring if the fasteners joining the mating surfaces are overtightened. O-rings can leak for many of the same reasons that induce a gasket leak, but O-rings also have their own unique failure causes. Potential O-ring failure causes include:

- As mentioned earlier, an O-ring fits into an O-ring channel formed by the mating parts. Incorrect O-ring channel dimensions can result in either too much or too little squeeze on the O-ring. Either condition can induce a leak. The O-ring channel may not have been machined to its correct drawing dimensions. (The failure analysis team should inspect the channel to ensure dimensional conformance.) The O-ring channel may meet drawing requirements, but the drawing dimensions may be inappropriate. O-ring manufacturers define the maximum and minimum O-ring channel dimensions for various sized O-rings, so the failure analysis team should compare these to the maximum and minimum dimensions allowed by the drawing.
- O-rings can be torn or twisted during installation. The failure analysis team should check for this condition when disassembling and inspecting leaking equipment.
- Some O-rings require backup rings, while others do not. A backup ring is designed to support the O-ring in the O-ring groove. Absence of a backup ring when one is required can induce leakage.
- As described previously for gaskets, the correct O-ring material should be used. The failure analysis team should have the O-ring material analyzed and confirm with the manufacturer that the design specifies the

correct material for the intended application. This pertains to the fluids being sealed and the pressure and temperature environment.*

- The O-ring could be the wrong size. The failure analysis team should inspect the O-ring to ensure that it is the correct part number and that it is dimensionally conforming.
- Contamination on the O-ring or in the O-ring channel can create a leak path, so the failure analysis team should look for contaminants when disassembling and inspecting an O-ring leak.
- As described previously for gaskets, corrosion can create a leak path. The failure analysis team should look for corrosion in the O-ring groove and on the mating surfaces.

Bond-Joint Leaks

Bond joints can be formed from epoxies, adhesives, or other sealants (such as room-temperature vulcanizing sealant). Potential epoxy-, adhesive-, and other sealant-leak causes include:

- Epoxies can be delivered in either a premixed, refrigerated configuration or in a two-part, unrefrigerated condition. Premixed epoxies must be kept at the proper temperature before use. (If the epoxy is not kept refrigerated, it will partially or completely cure prior to application.) Epoxies that require mixing prior to use may be mixed in an improper ratio, or the mixing may not be complete. Either situation could prevent complete curing and allow leakage. Sometimes, an incomplete cure can be determined by probing the epoxy joint for soft areas. If this is not conclusive, the failure analysis team can review work instructions for epoxy mixing and storage practices to assess if any anomalies occurred. The materials analysis technologies described in Chapter 11 can sometimes determine if the epoxy mix ratios were correct, although this is a relatively difficult determination to make after the epoxy has cured.
- If a two-part epoxy is mixed too aggressively, bubbles can form and induce porosity, resulting in a leak path. The failure analysis team can section the epoxy to determine if porosity is present. Sometimes, porosity can be seen with a simple external visual examination without sectioning the suspect area. N-rays may also reveal porosity in an epoxy joint.
- Most epoxies and many adhesives have a shelf life, which refers to how old the material can be before it should no longer be used. If the

*The *Challenger* space shuttle is perhaps the most famous example of using an O-ring outside of its intended environment. In this case, the O-rings sealing the solid rocket motor chambers were subjected to temperatures outside an acceptable range. The O-rings were too hard at low temperature and did not deform sufficiently to create a seal. Burning gases escaped and ignited the liquid propellant.

material shelf life has expired, the material may become brittle, not cure, or develop other conditions that create a leak path.

- Epoxies have a pot life, which refers to the window of time the operator has to use the epoxy after it has been mixed or after a refrigerated epoxy has been removed from refrigeration. Exceeding the pot life can similarly induce leak-creating conditions.
- Most epoxies and many adhesives have a specified cure time, which refers to how long the joined surfaces should remain immobile prior to being loaded. Cure times are a function of temperature and perhaps other environmental conditions. Subjecting the joined surfaces to loads prior to the cure time can induce either an immediate leak or a latent defect (i.e., one that appears later). Such a condition is difficult to discern from an examination of the failed hardware, but the failure analysis team may be able to confirm its existence by observing production practices.
- Epoxies and adhesives can be sensitive to surface-finish conditions. Assuming a clean surface (contaminants are discussed subsequently), a rougher surface generally provides a better seal than a smooth surface. If surfaces to be sealed are too smooth, the epoxy or adhesive may not form a structurally sound seal. Most adhesive and epoxy containers do not specify a maximum allowable surface finish (i.e., how smooth the surface can be), but the failure analysis team can usually obtain this information from the epoxy or adhesive manufacturer.
- Contaminants of any type on mating surfaces during epoxy or adhesive application decrease the likelihood of a good seal. Oils, silicones (e.g., WD-40), dust, and even skin oils can degrade a bond joint. If the epoxy or adhesive makes a clean break from one or both of the sealed surfaces, this usually indicates the presence of a contaminant. Fourier transform infrared spectroscopy is a good technology for identifying the presence of such contaminants.
- The epoxy or adhesive application process can be critical. For example, some manufacturers require wetting both surfaces to be sealed with the epoxy or adhesive and then pushing the two surfaces against each other. A common tendency is to wet only one surface with the sealant and then push the other surface into it. Surprisingly, doing this with sealants that require wetting both surfaces can result in a weak joint and subsequent leak.
- In some cases, operators apply too little epoxy or adhesive. When this occurs, the area with light or no epoxy or sealant application can leak.
- Corrosion on the mating surfaces (either galvanic or chemical) can create a leak path. Corrosion is usually evident under magnification.
- Either the surfaces or the fluids being sealed may be chemically incompatible with the epoxy or adhesive being used, or the epoxy or adhesive may be inappropriate for the temperature or pressure environment. The failure analysis team can check with the epoxy or adhesive manufacturer to confirm the use of an appropriate sealant.

Weld Leaks

Pressure vessels and other tanks often have weld joints, which provide a structural connection as well as a seal. Weld leak path causes include:

- The weld may crack if it is subjected to too much stress, if the wrong welding rod is used, or if the weld technique is poor. A weld crack generally appears as a visible line. If the line is not visible to the naked eye, it may become visible under magnification or by using a fluorescent or dye penetrant inspection technique, as discussed in Chapter 11. X-ray inspection can also be used to identify weld cracks.
- The weld may have a pinhole leak as a result of inadequate weld technique. (The weld may have been too hot or too cold, or the weld application speed may have been too high.) Fluorescent or dye penetrant inspections are useful to isolate pinhole leaks through welds.
- The weld may contain porosity induced by poor weld technique or contaminants. When subjected to the heat of welding, contaminants may turn into a gas, inducing porosity in the weld. Fluorescent or dye penetrant inspections can be used to identify the existence of a leak path. Porosity is visible in a weld x-ray.
- The weld may leak if too little weld material is used. Again, fluorescent or dye penetrant inspections can be used to identify the existence of a leak path.

Polyvinyl Chloride Leaks

Polyvinyl chloride (PVC) piping is widely used in a variety of applications due to the relatively low cost and ease with which PVC piping systems can be constructed. The PVC pipes and joints are typically solvent cemented. Leaks in PVC joints can be induced by the following causes:

- There are various types of PVC, and each requires a specific solvent and primer. If the wrong solvent or primer is used, the joint is likely to fail.
- Proper PVC joining requires the application of a primer to both surfaces to be joined, followed by the solvent application. The primer should be liberally applied until it becomes sticky (as evidenced by increased drag on the applicator). At this point, the solvent cement should be applied to both surfaces. If any of these steps are not performed correctly, the joint may leak.
- Proper PVC joining calls for removing the glaze on the surfaces to be joined with sandpaper to provide good solvent and primer penetration. If the glaze is not removed, the joint may leak.
- The PVC piping should be cut square and fully inserted into the joint to which it is being cemented. If the cut on the end of the pipe is not

square or the pipe is not fully inserted into its mating fitting, the joint may leak.

- As is the case with epoxies and other adhesives, solvent-cemented PVC joints have a specified cure time. If the joint is subjected to a load prior to the cure time, the joint may fail and leak.

In many cases, people who work with PVC joints often have not been adequately trained in how to properly apply primer and solvent cement. Observing the process is usually helpful in finding the cause of process-driven PVC joint failures.

Valve Leaks

Valves are used to seal containers or to control operating logic in hydraulic or pneumatic systems. Potential causes of valve leaks include:

- The valve may not have completely closed. Improperly adjusted valves, operator error, calibration errors, or contaminants in the valve seating area can keep valves open.
- The contaminants that can keep valves open may be carried to the valve seating area by the fluid being transported through the valve. Such contaminants can also be an accumulation of deposits resulting from chemical or mechanical interactions in the valve seating area.
- If valves are overtightened, the valve seating area can be damaged. If this occurs, the seating surfaces may no longer seal.
- Valves can leak if they are exposed to temperatures, pressures, or pressure spikes outside their design limits.
- Valves can leak if internal materials are not compatible with the fluids flowing through the valve.
- Valves can experience internal corrosion, such that the seating surfaces no longer seal.
- Valves can wear out, such that the seating surfaces no longer seal.
- Valves can dry out, resulting in the seating surfaces cracking so that they no longer seal.

Structural Leaks

Leaks can occur through the walls and lids of containers and pressure vessels. The causes of these types of structural leaks include:

- Porosity can provide a leak path. Incorrect weld technique or the presence of contaminants during welding can induce porosity. If pressurized containers are molded or cast, contaminants present when molding or casting can induce porosity. If the molding or casting is not done at the correct temperatures and flow rate, porosity can result. If the mold

gates (where molten material flows into the mold and where air escapes) are poorly located, porosity can result.
- Cracks in vessels and containers can provide a leak path. Cracks can occur as the result of the vessel or container being subjected to loads outside design limits, excessive vibration or shock, incorrect weld technique, operating the system outside its intended environment, or thermal shock. If all valves on a vessel are closed when it contains water and the temperature drops below freezing, the forces and pressures created may be sufficient to crack the structure.
- Corrosion can induce a leak path if the corrosion is deep enough to weaken the container or vessel.
- Chemical incompatibility can eat through a container or vessel wall if the structure is not compatible with the fluid being contained.
- Foreign object damage can puncture the structure if the container encounters the foreign object with sufficient velocity or force.

Structural leaks can usually be confirmed by direct observation of the structure for leaking material, holes, cracks, porosity, or other breaches of the container structural integrity. X-rays or n-rays (depending on the structure material) can be used to find porosity and cracks. If the item is small enough, it can be pressurized with air and submerged. (Bubbles show the leak path.) Dye and fluorescent penetrant inspections are useful for showing cracks, as explained earlier. On ferrous materials, magnetic particle inspection (also know as magnafluxing) can help reveal the outline of a crack. Magnetic particle inspection involves sprinkling iron particles around a suspect crack area and applying an electromagnetic field to the structure. If a crack is present, the iron filings outline the crack when the magnetic field is applied.

Summary

Many system failures involve leaks. Various sealing techniques are susceptible to different leak causes. Gaskets and O-ring leakage can be induced by material, configuration, and mating surface factors. Bond-joint leakage can be induced by epoxy or adhesive process anomalies and mating surface conditions. Weld leakage can be induced by material and weld technique issues. The PVC joints are susceptible to inappropriate solvent or primer use, process anomalies, and other factors. Valve leaks can be induced by any of several causes influencing the valve seating surfaces. Structural failures can be induced by porosity, cracks, foreign object damage, and other factors. All of the aforementioned are susceptible to contaminants, design, material compatibility, and environmental factors.

Example for Group Discussion. In the 1950s and early 1960s, the two best-selling motorcycles in the United States were Harley-Davidson (manufactured in the United States) and Triumph (manufactured in Great

Britain). Both motorcycles had a well-deserved reputation for leaking oil. When Honda started selling motorcycles to the United States in the mid-1960s, their motorcycles did not leak. Up until the time of Honda's arrival, the leaks that characterized Harley-Davidson and Triumph were not perceived to be a significant problem by most motorcyclists. After Honda arrived, however, the previously acceptable Harley and Triumph leaks became a problem for these two manufacturers. Harley-Davidson and Triumph sales declined sharply. Harley-Davidson nearly went out of business; Triumph did.

When Harley-Davidson embarked on its successful program to improve the quality of its motorcycles in the 1980s, one of the company's top priorities was to fix the oil leaks inherent to its motorcycles. When the Triumph brand was resurrected in England in the early 1990s, building motorcycles that did not leak was a key objective.

Why was freedom from leakage such a high priority for Harley-Davidson and Triumph?

Are the products offered by your company susceptible to leaks? How do you control this failure mode?

SELECTED REFERENCES

- P.C. Reid, *Well Made in America: Lessons from Harley-Davidson on Being the Best,* McGraw Hill Publishing Company, 1990
- *The APG O-Ring Handbook,* APG, 2007

CHAPTER **14**

Contaminants

CONTAMINATION has been mentioned in this book. Contaminants can be a cause of numerous types of system failures:

- Contaminants can act as an abrasive, accelerating wear on mating parts.
- Contaminants can act as a source of mechanical interference, preventing hydraulic actuators from working, valves from fully closing, or many other types of failures due to inhibited motion.
- Contaminants can degrade bond joints and welds.
- Contaminants can interfere with switches closing properly.
- Contaminants can degrade energetic material performance (e.g., fuels and explosives).
- Contaminants can interfere with painting and other coating operations, allowing corrosion to accelerate, coatings to flake off, and so on.

There are numerous techniques for confirming contaminant presence. Magnification, optical microscopy, and scanning electron microscopy can reveal the presence of contaminants. Any of the materials analysis technologies (energy-dispersive analysis of x-rays, Fourier transform infrared spectroscopy, spectrometry, chromatography, secondary ion mass spectrometry, and Auger) can be used to identify elements and, in some cases, compounds. Without confirmation by any of the preceding approaches, contaminant presence may sometimes be assumed based on material performance.

When the presence of a contaminant is suspected, the failure analysis team must find and eliminate the contaminant source, which can be obvious or quite subtle. The following list summarizes a few commonly encountered contaminant sources:

- Packaging materials may contaminate products. This can occur even with new packaging materials if the packing material flakes, generates dust, exudes oil, or otherwise transfers substances from the packaging material to the product. This may occur, for example, with paper-based, cloth-based, or expanded foam packaging materials. In many cases, packaging materials are reused. This can transfer contaminants from the products previously contained to the products currently contained.

- Materials handling equipment (such as conveyors, carts, trays, and other similar items used for moving materials between locations) can be a contaminant source. Materials handling equipment can become contaminated from any number of sources and then transfer its contamination to the products it contacts.
- Work areas (including machines and workbenches) can become contaminated and pass the contamination on to other products.
- People can contaminate products with skin oils, perspiration, breath moisture, or contaminants from other sources when handling products. In cases where the potential for human-transferred contaminants is recognized, workers may be required to wear finger cots or gloves. Even with these protective measures, however, contaminants can still be transferred to the product if the worker touches a contaminant source while wearing the finger cots or gloves or if the finger cots or gloves are not changed frequently.
- Sometimes, contaminants can be airborne. This may occur if dust or paints are present.
- Some processes use etchants with a subsequent rinse. Very small-diameter wires are produced by exposing a larger-diameter wire to an acid etch to reduce the diameter and then rinsing the wire to remove the etchant after the desired diameter is attained. If any acid is left on the wire, the acid-etching process continues, and the wire diameter becomes too small. In any process where a rinse cycle is used and contamination is suspected, the failure analysis team should determine if the rinse completely removes whatever it is supposed to remove.
- Test solutions can contaminate products or result in an out-of-calibration condition if the test solution is contaminated. Chemical-concentration-monitoring equipment, for example, usually requires calibration with a test solution of known concentration. If the test solution is contaminated, it can result in inaccurate monitoring-equipment calibration.
- Cleaning solutions can contaminate products if the solution is contaminated.
- Lubricating oils can become contaminated and fail to provide their required lubricating function, or they can transfer contaminants into critical areas. Lubricating oils can be a contaminant by themselves if the oil migrates into an area where it should not be (e.g., onto electronic components or optical elements or areas to be bonded or welded).
- Cutting oils used for machining parts can interfere with subsequent operations (bonding, welding, painting, etc.) if the cutting oils are not completely removed after machining.

- Silicones are notorious for their ability to migrate, with subsequent adverse affects on electronic components, optics, surfaces to be joined, surfaces to be coated, and energetic materials. WD-40, a popular silicone-based spray lubricant, is ubiquitous in many manufacturing and maintenance organizations. If the failure analysis team is concerned about potential contaminant sources in any of the aforementioned or similar product areas, the presence of lubricants such as WD-40 strongly suggests that the suspected contaminated condition exists.
- Outgassing refers to the ability of a material to transfer from a solid directly into a gaseous state. The concern here is that material in the gaseous state subsequently becomes deposited elsewhere in the system. Many plastics outgas. The most common example is the film that forms on the inside of car windows. This occurs when the plastic components in an automobile interior outgas, and the gaseous plastic then deposits itself on the inside of the windshield. This is more of a nuisance than a problem, but in sealed systems, it can become a significant problem. In some systems, outgassing can actually insulate the contacts on an electrical switch.
- Chapter 13 discusses the adverse effects of contamination on epoxies and other sealants. Epoxies and sealants can be a contaminant in their own right if these materials migrate into other-than-intended areas. If an epoxy or a sealant flows into areas where there are moving parts, for example, the epoxy or sealant can interfere with the part motion.

The preceding list is intended to stimulate the reader's thinking about potential contaminant sources. It is not intended to be a complete list. When searching for contaminant sources, the failure analysis team should consider the aforementioned as well as other sources unique to the failure being analyzed.

Example—Titanium Component Washing

Litton Lasers machined lightweight titanium components using conventional machining equipment and cutting oils. In a subsequent operation, Litton cleaned the titanium parts by wiping them and then using ultrasonic cleaning equipment. The ultrasonic cleaner contained a solvent to dissolve and wash away the cutting oils. (Ultrasonic vibration helped the solvent penetrate into all areas on the titanium parts.) After ultrasonic cleaning, the parts were allowed to dry and then were epoxy-bonded to optical elements. After the epoxy cured, the subassemblies underwent shock and vibration testing to confirm bond adequacy.

Litton experienced frequent epoxy disbonds during the shock and vibration test cycles. The failure analysis team prepared a fault-tree analysis and a failure mode assessment and assignment. One of the hypothesized failure modes was epoxy disbonding due to the presence of contaminants on the bonding surface. The failure analysis team examined the failed bond joints and discovered that clean separations had occurred between the epoxy and the surface of the titanium components. This clean separation is a classic characteristic of a bond-joint failure due to contamination. In such situations, the epoxy separates cleanly from the part because contaminants prevent the epoxy from bonding to it.

With its understanding of the manufacturing process, the failure analysis team suspected that the ultrasonic cleaning operation was not completely removing the cutting oils from the contaminated parts. To address this concern, the failure analysis team used Fourier transform infrared (FTIR) analysis to assess cutting oil contaminant levels on the titanium parts before and after the ultrasonic cleaning operation. To the team's surprise, the contaminant levels on the titanium parts after ultrasonic cleaning were higher than they were before ultrasonic cleaning. Not only did the ultrasonic cleaner fail to completely remove the cutting oils, it was actually increasing the amount present on the parts.

The failure analysis team visited the ultrasonic cleaning equipment work area and asked the operator to demonstrate how the cleaning equipment worked. The operator ran through a cleaning cycle. He explained to the failure analysis team that the solvent had to be kept at the minimum depth indicated by a marker on the solvent reservoir. The operator explained that the solvent depth dropped approximately a quarter of an inch after each cleaning cycle, and that he added solvent after each cycle to maintain the required depth.

The failure analysis team peered into the solvent reservoir and observed that the solvent was a thick, dark liquid. Fresh solvent is clear.

The failure analysis team asked for the ultrasonic cleaner operating manual, which the operator did not have. The failure analysis team called the ultrasonic cleaner manufacturer, who informed the team that the solvent should be flushed and replaced daily. The operator explained that he had never completely drained and replaced the solvent; he just added to it as the solvent level dropped after each cleaning cycle.

The failure analysis team asked the operator to flush the solvent and clean the titanium parts again. After this occurred, the failure analysis team checked before-and-after component cleanliness with the FTIR. The team found no remaining oils on the part cleaned with fresh solvent. After implementing this change, the optics debonding failures dropped significantly.

Summary

Contaminants can cause accelerated wear on mating parts, mechanical interference, degraded bond joints and welds, undesired insulation on switch gear, degraded energetic material performance, interference with painting and other coating operations, and many other problems. Magnification and materials analysis techniques reveal contaminant presence, and contaminant presence may be inferred based on product performance. Contaminant sources include (but are not limited to) packaging, materials handling equipment, work areas, people, dust, etchants, cleaners, test solutions, lubricating oils, cutting oils, silicone, outgassing, and migrating epoxies and sealants.

Example for Group Discussion. In the titanium component washing case study, the operator continually added solvent to the ultrasonic cleaner to maintain the required solvent level. However, the ultrasonic cleaner equipment manufacturer recommended replacing the solvent on a daily basis.

When the operator continually added fresh solvent to the solvent reservoir (instead of replacing the solvent), what was the effect on the solvent?

What was your impression when you learned that the operator did not have an equipment manual for the ultrasonic cleaner? Who should have recognized and rectified this situation?

Could the ultrasonic cleaning equipment manufacturer have done anything differently to alert operators that the solvent should be replaced on a daily basis?

Will replacing the solvent on a daily basis always prevent contamination by the ultrasonic cleaner?

What should the titanium components manufacturer do to rectify the aforementioned situation?

SELECTED REFERENCES

- L.E. Murr, *What Every Engineer Should Know about Material and Component Failure, Failure Analysis, and Litigation,* Marcel Dekker, 1987
- J.P. Sibilia, *Materials Characterization and Chemical Analysis,* Wiley VCH, 1996

CHAPTER **15**

Design Analysis

IN SOME CASES, the failure analysis team finds that all components meet their requirements, the system was properly assembled, and it was not operated or tested in an out-of-specification manner, yet it still failed. When this occurs, the only conclusion the failure analysis team can reach is that it missed something in its analysis or that the design is defective. This chapter focuses on the latter possibility.

Based on the author's experience, many people (especially the system designers) may be reluctant to conclude that designs can be defective. This is perhaps a natural reaction and explains why a first reaction is to find a nonconforming part or to conclude that the system was operated incorrectly.

If one thinks about the design process, though, it is not too difficult to understand that engineers make mistakes. This can occur for any of several reasons:

- Engineers are human beings who sometimes make mistakes in their calculations, assumptions, and understanding of what the system is supposed to do.
- Engineers may not recognize the environmental extremes (both external and internal) that the system will experience.
- Performance, reliability, transportation, and other requirements may not be fully understood or taken into account during the design.
- Worst-case tolerance conditions (both mechanical and electrical) may not be fully analyzed during the design, which can result in excess clearance, mechanical interference, or degraded system performance.
- The requirements flowdown from the system level to the subsystem or component level may not appropriately define the performance, loads, environments, or other requirements the subassemblies and components must meet under worst-case system conditions.

When the situation described at the beginning of this chapter is encountered (i.e., all components meet their requirements, and the system was properly assembled and not operated in an out-of-specification manner), a systems failure analysis is really a continuation of the engineering development process. In this situation, a systems failure analysis further defines required actions to ensure acceptable system performance under all conditions. When the failure analysis team finds this to be the case, it is recommended they consider evaluating the areas described as follows.

Requirements Identification and Verification

The first step of any design is to define the requirements. This involves defining what it is the product is supposed to do, under what environments it is supposed to do it, and for how long it is supposed to do it. Based on a clear definition of these requirements, the engineering effort should develop designs that can be verified as having met all of the requirements. This verification usually occurs through analysis and test.

The system requirements are stated in a set of engineering drawings and sometimes in specifications. Government projects normally have comprehensive specification requirements. In many cases, commercial products do not have a specification defining performance, reliability, and other requirements. Engineering drawings typically define dimensional and material requirements but not performance, reliability, safety, and other requirements, which are typically defined in specifications. The failure analysis team should review with the engineers exactly what requirements the product is supposed to meet.

When the system requirements are clear (especially with regard to the requirements that were not met), the failure analysis team should evaluate how compliance with these requirements was verified during the engineering development process. Do analyses exist that show the design met the design requirements? Did tests verify that the system met its performance requirements? If tests were used, did the test strategy make sense and provide statistically significant results? Did the engineers use intuition or assumptions in concluding that the design met its requirements, or did they use quantitative and objective tests and analyses?

If sound analyses and meaningful tests confirm that the design met its requirements, the failure analysis team can conclude that the design is adequate. If this is not the case, additional tests or analyses are required to make this determination. In some cases, the additional tests or analyses identify where the design requires modification.

Requirements Flowdown. When the system-level requirements are understood, it is the engineers' responsibility to take the system-level requirements and flow them down to subassembly and component-level requirements. Engineers should use sound engineering practices and the laws of physics, materials science, and other accepted, technically correct approaches.

Consider the case of an electromechanical system that must operate in a 140 °F environment, which is a very common upper-end temperature requirement for such a system. The engineers must define the upper-level temperature specification for the components and circuit card assemblies inside the system. One may be tempted to simply say that these should be specified with an upper temperature requirement of 140 °F, too, but that would be wrong. The temperatures inside the system will be considerably higher. The engineer (using electrical circuit analyses, heat-transfer equations, and other engineering principles) must define the worst-case internal temperatures and apply these temperature requirements when procuring the electrical components and circuit card assemblies.

Similarly, engineers may design a hydraulic system to operate at 3000 psi. Suppose the system experiences extremely short pressure transients that can momentarily increase the pressure up to 3500 psi. Suppose further that the system requires pressure transducers to monitor system pressure and record any transients. If the engineers specify pressure transducers capable of withstanding 3000 psi with a very long sampling interval, the transducers will most likely miss the brief pressure spikes, and they will fail as a result of the pressure exceeding their maximum rated pressure.

In most situations, manufacturers buy from suppliers the types of components and subassemblies described earlier. If the engineer creates a subassembly specification stating his or her requirements, and the supplier provides a subassembly that meets the specification, then the supplier has met its requirements. If the supplier's subassembly fails because it experiences loads or environments exceeding the engineer's specification, the fault lies with the engineer who generated the specification (and not with the supplier). The requirements flowdown was inadequate. The engineer must understand the loads and environments and select appropriate components.

The reader may wonder how likely it is that an engineer would make the types of mistakes described earlier. In the author's experience, these types of mistakes are not uncommon. This is another reason why having a blend of disciplines (engineering, manufacturing, quality assurance, purchasing, etc.) on a failure analysis team makes sense.

Mechanical Interference or Excessive Clearance. If the failure analysis team hypothesizes that mechanical interference or excessive clearance could have induced the failure, the team should identify all components that may have created the condition. The team should have the components inspected to determine if they meet drawing requirements in areas that could have induced the interference or excess clearance. While this is occurring, the team should also request a tolerance analysis to show if the interference or excess clearance is allowed by the engineering drawing tolerances. A tolerance analysis shows if an interference condition (Fig. 15.1) or excess clearance exists.

In many cases, engineers perform design analyses for proper device operation using nominal dimensions. The engineers may not have considered worst-case tolerance conditions. Even if the engineer has performed

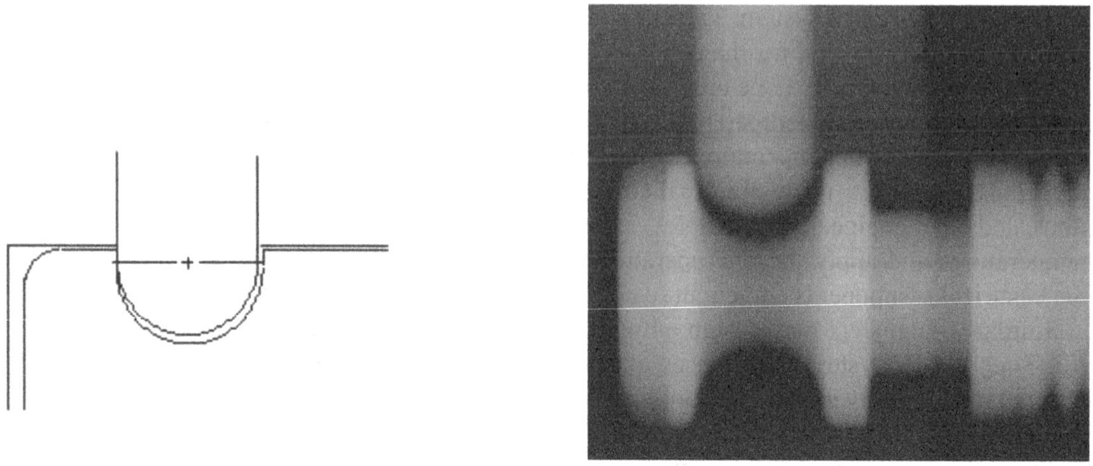

Fig. 15.1 Tolerance analysis and x-ray images. A mechanism designer and manufacturer experienced pin jamming in one of its products. A fault-tree analysis hypothesized the existence of an interference fit. The tolerance analysis drawing on the left showed that this could occur under certain tolerance conditions. An x-ray of the failed mechanism confirmed the failure cause.

a tolerance analysis, the worst-case scenario may not truly be the worst case. Defining the worst case involves an in-depth understanding of how the system operates, so that the tolerance analyst knows which dimensions to take to their maximums and which to take to their minimums. This is not an easy thing to do, and it is an area where mistakes are easily made.

This may be a surprise, but many engineers do not consider worst-case conditions when developing a design. In fact, it is usually not the engineers who determine nominal dimensions and assign tolerances to these nominal dimensions. The job of determining dimensions and tolerances is usually assigned to the computer-aided design (CAD) operator. In most companies, CAD operators do not analyze what is required to make the system work under worst-case tolerance conditions. The CAD operators typically select tolerances based on a set of standardized tolerances for their industry. These standard tolerances can be based on the nominal dimension, the component production method, the number of decimal places to the right of the decimal point, or other such schemes.

In most situations, tolerances are not based on what is required to ensure acceptable system performance under all conditions; they are instead based on a standard scheme unrelated to system performance. Under one tolerancing scheme, parts up to 1 in. may have a standard tolerance of 0.002 in., up to 5 in. may have a tolerance of 0.005 in., and so on. Under another tolerancing approach, the CAD operator may assign a tolerance of 0.005 in. to all machined parts, 0.010 in. to all cast parts, and so on. Under yet another tolerancing approach, the CAD operator may assign a tolerance of 0.003 in. to a nominal dimension of 3.125 in. because there are three places to the right of the decimal point, 0.010 in. to a nominal dimension of 3.25 in. because there are two places to the right of the decimal point, and so on.

Circuit Performance

Some failure analyses involve electrical circuit evaluations. Engineers can analyze circuits in a manner similar to mechanical assemblies. Electrical and electronic circuits can be modeled with an equation. When the equation has been written, the engineer can specify input values and calculate output values. The circuit has electrical and electronic components, and, just like mechanical design, these components have tolerances around their nominal values for resistance, capacitance, and so on. There are several approaches that engineers can take to analyze circuit performance, particularly under worst-case tolerance conditions:

- The engineer can use a commercial circuit analysis program, such as PSpice, to analyze circuit performance.
- The engineer can take each component to its worst-case value and use these values in the circuit equation to determine if the circuit output is acceptable. However, this is not always a good approach, for the same reasons outlined for mechanical components. The engineer must make assumptions about which end of each tolerance band constitutes the worst case.
- The engineer can use Excel Visual Basic programming capabilities to run a Monte Carlo circuit simulation. This involves using the circuit equation, the nominal value and tolerances of each component, the nominal value and tolerances of each input, and the acceptable circuit output range. The engineer can place the circuit equation in a loop and randomly vary the value of the inputs and each component (within their specified tolerances) for tens of thousands or even millions of loops. The program can be written to calculate the output value for each set of randomly assigned input and component values, and the circuit output can be assessed to determine if it lies within the acceptable range. At the end of the multiple simulation runs, the program can be set up to calculate the percent of acceptable versus unacceptable output values. This models how often the circuit provides an acceptable output.

In addition to the aforementioned, circuit designers can assess the temperature experienced by each component under all circuit operating conditions. Component performance and life expectancy can be significantly degraded as temperature increases. If the failure analysis team finds that this has not been done, and it suspects or finds that the temperature induced an electrical component failure, the team should ask for this analysis.

In circuit failure analysis situations, the failure analysis team should ask how the engineer evaluated circuit performance under all tolerance and temperature conditions. If the failure analysis team finds the engineering development effort did not exercise all component and input tolerances to ensure acceptable performance under all conditions, and no other failure causes are discovered, the team should request the circuit analyses described previously.

Chapter 3 made a reference to sneak-circuit analysis as a cause of inadvertent function. The example was one that occurred in the 1950s, when an automobile manufacturer produced a car with a sneak circuit that activated the windshield wipers when the radio was operated. A sneak circuit is a design mistake. It sends power or a signal to a location it is not supposed to. Sneak-circuit analysis is a specialty capability offered by only a few suppliers, and it is expensive, but it is not really necessary if the failure analysis team uses the fault-tree analysis technique to identify all potential failure causes. If a sneak circuit is present, the fault tree will find it.

Mechanical Failures

Chapter 12 describes the analysis of broken mechanical components to determine if they failed as a result of fatigue, brittleness, excessive load, and so on. When the failure analysis team makes this determination and confirms that the failed component conforms to the drawing, was fabricated from the correct material, did not experience any out-of-specification loads, and so on, it should assess if the component design is adequate.

During the design process, the intent is usually to make the component stronger than the loads it will experience, thereby assuring it will not fail. To do this, development engineers analyze loads and convert them into stresses. The development engineers then determine the maximum stress the component material can withstand and design the component such that the stress levels are substantially lower than the component strength. The concept is to have an adequate difference between the load and the component strength.

A typical approach for doing so is to design the component so that its strength is 1.5 times greater than the maximum anticipated load. The 1.5 factor is referred to as a safety factor. In some applications, the safety factor can be higher. For other applications (especially where weight is a consideration), the safety factor may be lower (e.g., on high-performance aircraft). In some cases, instead of simply making the component strength 1.5 times greater than the maximum anticipated load, development engineers characterize the load and component strength distribution. After defining the means and standard deviations of each distribution, development engineers can then design the component such that its strength is several standard deviations higher than the load.

For components that will be subjected to repeated stress reversals, engineers consult material-specific stress-number of cycles (*S-N*) diagrams to ensure that the component will have an acceptably long life. The *S-N* diagram plots stress-reversal magnitude levels against the number of stress-reversal cycles. As the stress-reversal magnitude increases, the part fails with fewer stress-reversal cycles. Below a certain stress-reversal level, the part has infinite life. Ideally, engineers should design parts subjected to stress reversals to be in the infinite life range of the *S-N* curve.

If the failed component conformed to the engineering drawing, was made of the correct material, experienced no out-of-specification loads, and so on, the failure analysis team should ask if the engineer prepared the aforementioned analyses during the component engineering development. If not, the part may have failed because its design was too weak for its intended use. The failure analysis team should review all assumptions associated with the engineering development stress and fatigue life analyses and factor in worst-case tolerance conditions. In some cases, the component may be strong enough at nominal dimensions but weak enough to fail at worst-case tolerance conditions. This can sometimes be true when considering the component in isolation, and it can sometimes be true when considering relationships influenced by other component tolerance stack-ups (e.g., adhesive bond widths, fastener alignment, etc.).

Materials Compatibility

In some cases, materials incompatibility can cause failures. This can occur as the result of galvanic corrosion or as a result of differences in thermal expansion as temperatures change.

Whenever corrosion occurs between materials in direct contact, the failure analysis team can consult specifications showing compatibilities between various materials. If these sources show that the mating materials are incompatible, the design is inadequate. Either the materials should be changed, or a coating should be applied to eliminate the direct contact.

Any time a fracture, warping, or leaking occurs between components made of dissimilar materials, the failure analysis team should evaluate the coefficients of thermal expansion for both materials. The different coefficients of thermal expansion determine how the dimensions of each component change with temperature. If the dimensional difference between two joined components is large enough to induce the distortion, the design is inadequate. Different materials or joining techniques are required to eliminate or reduce the distortion.

Environmental Factors

If a failed component conforms in all areas (yet still failed), the failure analysis team should analyze the environment in which the component operated. Several possibilities exist:

- The environment was outside the specified operating environment of the system.
- Storage conditions created an environment that was outside the system storage environment. This is frequently true for containers left in direct

sunlight, because the internal temperature of the container is significantly higher than the ambient external temperature.

- Conditions inside the system created an environment outside the range of individual component or subassembly allowed environments. As mentioned earlier, electronic components inside a structure can experience substantially increased temperatures.

The first step when considering if the environment played a role in the failure is to ask how the development engineers factored environmental considerations into the system design. In addition to considering temperature, the failure analysis team should also determine if temperature cycling, shock, vibration, and humidity were considered in the development process. If not, the failed item may not have been adequately designed to withstand the environments it experienced.

Examples

The following examples illustrate the value of design analysis.

Springs. Whenever the hypothesized failure causes include a spring, the failure analysis team should carefully scrutinize both the spring design and its conformance to design requirements. It is usually not too difficult to manufacture a spring to meet its dimensional requirements, but it is difficult to attain spring force-versus-deflection consistency. Differences in force versus deflection can vary widely from spring to spring, and for this reason, it is not uncommon to see spring designs with ±20% tolerances on their force values. This suggests two possibilities as suspected failure causes:

- Does the spring meet its force requirements? The failure analysis team should recognize that even though the springs were accepted, they were probably not inspected on a 100% basis. This, coupled with the difficulty of attaining spring-to-spring consistency, strongly suggests that individual springs may not have met their drawing requirements.
- Did the development engineers consider the worst-case force tolerances of the spring when designing the system, or did the development engineers assume the nominal value of the spring?

If the answer to the second question is that nominal spring values were assumed during development, the failure analysis team should pursue additional testing and analysis to evaluate spring performance at tolerance extremes.

Apache Main Rotor Blade. Chapter 1 mentions the Apache helicopter main rotor blade failures. McDonnell Douglas Helicopter Company engineers designed the Apache helicopter blade (a bonded stainless steel, carbon fiber, and titanium structure) to withstand a direct hit from high-explosive enemy projectiles. The design used four longitudinal spars that ran the length of the blade to provide structural redundancy in the event of a direct hit from an enemy projectile. The blade met its survivability requirement, but it did not meet its service life requirement. The Apache blade was supposed to have a life of 2000 flight hours. The Army found that the average blade life was approximately 800 hours.

Composite Structures built the Apache helicopter blades to a McDonnell Douglas design. The stainless steel rotor blade spars were bonded together with an epoxy seal. During the first 10 years of production, Composite Structures experienced a high rotor blade failure rate. The failure mode causing the in-house Composite Structures blade rejections was an epoxy disbond between the blade spars. When the blades failed in the field, they exhibited the same failure mode.

Both Composite Structures and McDonnell Douglas intermittently attempted to correct the blade disbond problem during the first 10 years of production. Both organizations suspected that contamination was the failure cause and focused their attention in this area. Composite Structures produced the blades in a clean-room environment and carefully monitored it for potential contaminants, but no contaminants were found in the disbonded blades. The problem persisted.

Composite Structures assembled a failure analysis team to focus on this problem. The team prepared a fault-tree analysis and investigated numerous potential failure causes. The fault-tree analysis identified inadequate bondline dimensional control as a potential failure cause. The team found that the bondline dimensions were not directly dimensioned by the rotor blade drawings but were instead the result of dimensions on the mating stainless steel spars.

The Composite Structures failure analysis team asked McDonnell Douglas what the minimum bondline dimension needed to be to ensure structural integrity. Engineers from McDonnell Douglas told the failure analysis team that the blade needed a minimum bondline width of 0.440 in. Composite Structures quickly prepared a tolerance analysis that showed the McDonnell Douglas design allowed the bondline width to be as low as 0.330 in. under worst-case conditions.

The Composite Structures team found it could measure the rotor blade bondline width (using x-ray inspections) on disbonded blades. The Composite Structures team found that all of the failed blades were near the worst-case bondline width of 0.330 in. None of the blades above 0.440 in. had disbonded.

Composite Structures tightened the tolerances on their manufacturing drawings to keep the minimum bondline width above 0.440 in. After doing this, both the in-house and field failures decreased dramatically.

Summary

If a system is properly assembled using conforming components and operated within its specified environment yet still fails, the failure analysis team should suspect the design. Design shortfalls usually result when development engineers have not fully considered operating and storage environments, worst-case tolerance conditions, worst-case loads, material incompatibilities, and other factors outlined in this chapter.

When all of the parts and assembly operations associated with hypothesized failure causes are found to be conforming, the failure analysis team should revisit the engineering development effort. The failure analysis team should look for areas that the development engineers may not have considered to determine if the design is adequate. Where design shortfalls are discovered, the failure analysis team and the development engineers should develop corrective actions to address the shortfalls.

Example for Group Discussion. What checks and balances should be incorporated during the engineering development process to ensure all environments, loads, and other factors are considered?

What difficulties should the failure analysis team anticipate when evaluating a failed system design?

In the Apache main rotor blade situation described in this chapter, why did prior failure analyses not uncover the failure cause? Why do you think the Army, McDonnell Douglas, and Composite Structures tolerated the high failure rate? What could these organizations have done to fix the problem earlier?

SELECTED REFERENCES

- B.R. Fischer, *Mechanical Tolerance Stackup and Analysis,* Marcel Dekker, 2004
- C.J. Monier, *Electric Circuit Analysis,* Prentice-Hall, Inc., 2001

CHAPTER 16

Statistics and Probability

FAILURE ANALYSIS can sometimes involve considerations of statistics and probability. This is especially true if the system fails even with adequate compliance to design (that is, all components are found to be conforming, the system was properly assembled, and the system operated in its intended environment). For this reason, this chapter can be considered as an extension of the preceding chapter on design analysis.

Statistics is a branch of mathematics dealing with the collection, analysis, and interpretation of data from a sample population of objects or measurements. Statistics refers to gathering numerical data to present information about a subject. Probability is another branch of mathematics that involves using these data to make inferences about the likelihood of an event occurring.

Statistical Distributions

Statistical distributions are mathematical functions that define the probabilities of outcomes from samples taken from a population or set of data or objects. Before any set of data can be analyzed statistically, one must select a statistical distribution that will be used to model the data in terms of the likelihood (or probability) of outcomes from sampling the population of data or objects.

The various types and selection of statistical distributions are beyond the scope of this chapter. However, it is useful to review some of the basic types of statistical distributions in order to understand some basic principles in their use. The main focus is on the uses of the normal distribution, which is the most commonly used statistical distribution. When the choice of a statistical distribution is not straightforward, statistical tests must be done to determine whether the chosen distribution provides an adequate fit to the data. These statistical tests—referred to as "goodness-of-fit tests"—are described in most introductory books on statistical analysis.

Discrete Statistical Distributions

When a population or data set consists of discrete values or objects, then discrete statistical distributions are used to analyze the likelihood of outcomes when samples are taken from the population. Three basic types of discrete statistical distributions include:

- Hypergeometric distribution
- Binomial distribution
- Poisson distribution

The hypergeometric distribution is a very basic type of probability distribution, whereby samples are taken from the population without replacement. That is, samples are not placed back into the population after sampling. In this case, the probability of outcomes is conditional, because the sampling process changes the population distribution for subsequent samples. The hypergeometric distribution is used when small populations are sampled without replacing the samples back into the population.

The binomial distribution is a discrete probability function that is based on sampling with replacement. The binomial theory is only correct when samples are returned back into the population. However, when the size of the sample (n) is small in comparison to the number of objects in the overall population or data set, the binomial distribution is an excellent approximation, even if the population is sampled without replacement.

The binomial distribution is frequently used to model the number of defective items in samples drawn from large lots of items. The two key parameters of the probability distribution function are the sample size (n) and the proportion (p) of defective parts in the lot or population. Thus, the probability (P) of x number of defective parts in a sample size (n) is:

$$P(x) = \frac{n!}{x!(n-x)!} p^x (1-p)^{(n-x)}$$

With a binomial distribution, the mean of the distribution is np, while the variance of the distribution is npq, where q is the proportion of nondefective parts (i.e., $q = 1 - p$).

For example, consider a sample size of five ($n = 5$) taken from a large lot of parts. If the design or manufacturing requirement specifies that 90% of the parts in the lot meet or exceed a certain mechanical property, then the proportion (p) of defective parts in the lot would be no more than 10%. In this case, Fig. 16.1 plots the probability of occurrence (P) for selecting x number of substandard parts from a sample size of 5 ($n = 5$) from a population containing 90% good parts ($p = 0.10$).

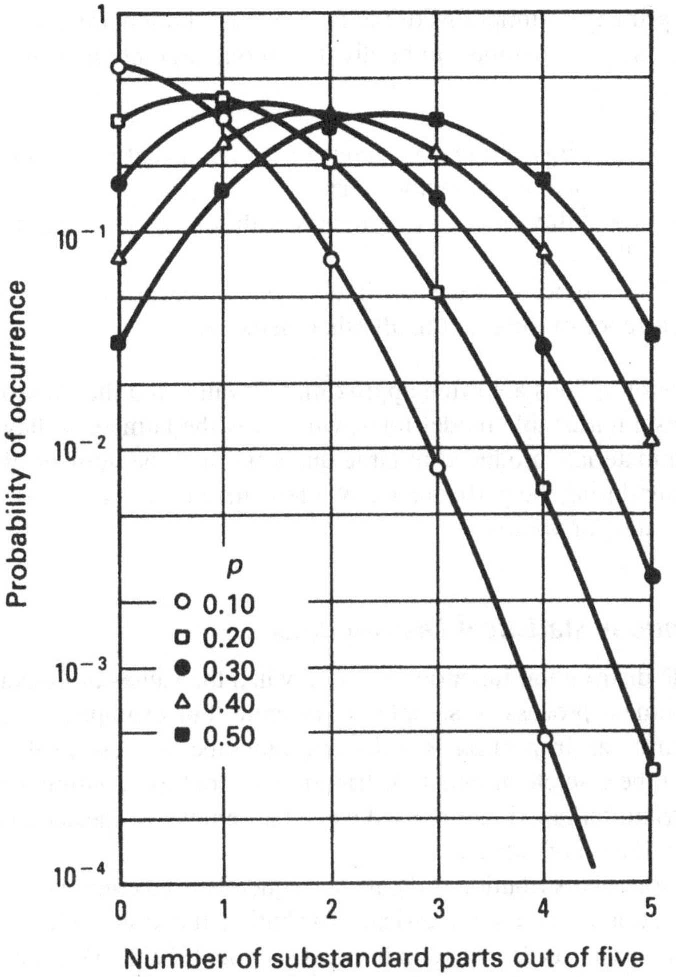

Fig. 16.1 Probability (*P*) of occurrence of *x* number of substandard parts when sampling five parts ($n = 5$) from a large population containing *p*% defective parts. Lines do not imply a continuous function.

As the percentage of good parts in the tested population decreases, the likelihood of seeing a high percentage of substandard parts increases. Note that for a *p*-value of 0.50, the probability plot is symmetric. For high *p*-values, the greatest probability of occurrence tends toward a low number of substandard parts.

The Poisson distribution is an approximation of the binomial distribution when the sample size (n) is large but when the outcome is a relatively small proportion (p) of the overall data set, such that $np < 5$. The Poisson distribution is used to model the occurrence of isolated events in a continuum; that is, rare events occurring in a given period of time, area, or volume.

The primary assumptions of the Poisson distribution are that the period of time, area, or volume can be divided into a large number of elements, such that:

- The probability of the occurrence of the rare event in a given element is proportional to the element size.
- The probability of two occurrences within a single element is infinitesimally small.
- The occurrence of the event in a given element is independent of the occurrence of the event in all other elements.

These assumptions are often approximately valid, and the Poisson model provides a reasonable model for x, where x is the number of flaws or defects in materials produced in large quantities and the number of failures occurring during the performance of a repetitive task in a specified period of time, area, or volume.

Continuous Statistical Distributions

Discrete distribution functions are used when the values or outcomes of a measurement process or sample are discrete. For example, the pass-fail criterion of an inspection is a discrete outcome, as described earlier in terms of the discrete binomial distribution. In contrast, continuous statistical distribution functions are used when the samples or measurements can be a continuum of values.

The normal distribution is the most frequently used continuous distribution function. It is a symmetrical distribution function centered about a mean value (μ), as described in the next section. Like the Poisson distribution, the normal distribution also can be considered as an approximation of the binomial distribution (Fig. 16.2). Both the Poisson distribution and the normal distribution can be used when the sample size (n) is large, which explains the ease with which the Poisson and normal distributions are used in many cases.

However, the normal distribution also can be a suitable approximation with a small sample size when the population is approximately centered about some given value (that is, $p \rightarrow 0.5$). For example, consider the earlier example (Fig. 16.1), where the probabilities of outcomes (P) are plotted for a relatively small sample size ($n = 5$) with a binomial distribution. If the population is 50% defective ($p = 0.5$), then the probability curve in Fig. 16.1 becomes symmetrical, with $p = 0.5$. This symmetry of the probable outcomes represents the condition whereby a normal distribution could be reasonable, even though the sample size (n) is small ($n = 5$). If more samples are taken ($n > 50$), then the approximation of the normal distribution is reasonable when the population proportion is between 0.20 and 0.80. Otherwise, the approximation of a normal distribution ceases to

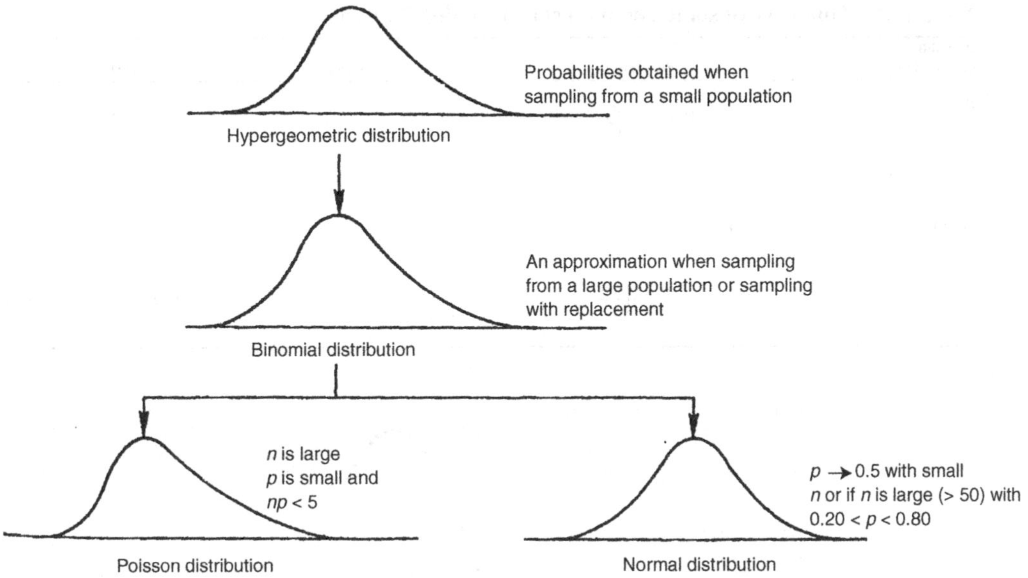

Fig. 16.2 Poisson and normal distributions as approximations of the binomial distribution, depending on sample size (n) and the proportion (p) of values in a large population

be reasonable. Like the binomial distribution, the normal distribution also is based on the assumption of either sampling with replacement or taking only a relatively small number of samples from a much larger population.

Besides the normal distribution, other forms of continuous distribution functions include:

- The log normal distribution, which applies when zero probability must be assigned to negative values of a quantity (e.g., lifetime of a component)
- Weibull distributions, which can be centered about a mean (μ) similar to a normal distribution or can be slightly skewed (positively or negatively) about the mean. The Weibull function includes a parameter for population size and a parameter for population shape.
- The exponential distribution, which is a positively skewed distribution that provides some analytical simplicity (Table 16.1). It is a special case of the more complex Weibull distribution.

Normal Distribution

To more fully understand how the preceding concepts can impact a failure analysis, consider the normal distribution. The normal curve (so named because it describes the statistical distribution of large populations or data sets in the world) has the typical appearance shown in Fig. 16.3.

Table 16.1 Summary of some common statistical distributions

Probability distribution	Function	Mean	Variance
Poisson	$P(x)=\dfrac{\mu^{x}e^{-\mu}}{x!}$	μ	μ
Exponential	$f(x)=\dfrac{1}{a}e^{-x/a}$	α	α^2
Normal	$f(x)=\dfrac{1}{\sigma\sqrt{2\pi}}e^{-(x-\mu)^2/2\sigma^2}$	μ	σ^2
Log normal	$f(x)=\dfrac{1}{\sqrt{2\pi}\,\sigma x}\ \exp\left[\dfrac{1}{2}\left(\dfrac{\ln(x)-\mu}{\sigma}\right)^2\right]\quad x>0$	$\exp\left(\mu+\dfrac{\sigma^2}{2}\right)$	$[\exp(\sigma^2)-1]\ \exp(2\mu+\sigma^2)$

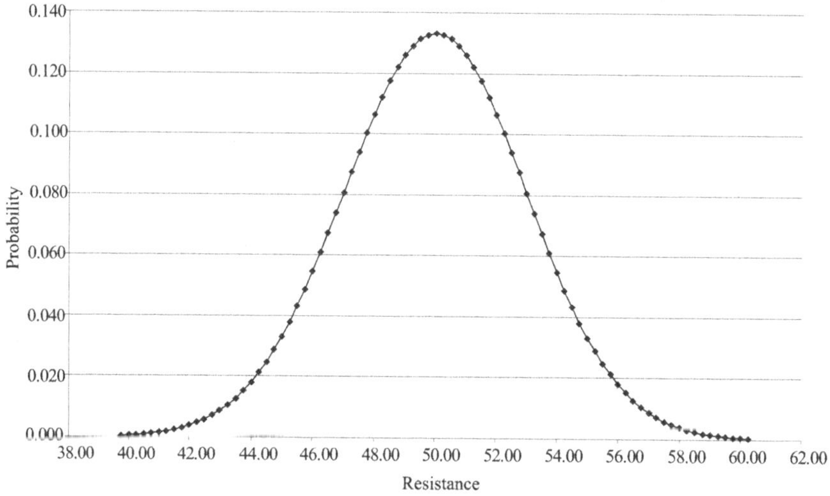

Fig. 16.3 Normal distribution curve showing the distribution of values about the mean for resistor resistance. The mean value for this distribution is 50 Ω. The area under the curve is equal to 1.

The normal curve is described mathematically as:

$$y=\frac{e^{-(1/2)(x-\mu)^2}}{\sigma\sqrt{2\pi}}$$

where y is the number of times a value occurs in the distribution; e is 2.7182, which is a constant that is the base of the natural logarithm; x is the value whose frequency is to be found; μ is the mean; σ is the standard deviation; and π is the ratio of the circumference of a circle to its diameter, ≅22/7 ≅ 3.1416, which is another constant.

In the previous equation, all of the terms except two are constants. The two factors that are not constants are the mean value (which is represented by μ) and the standard deviation (which is represented by σ). The symbol μ is the mean (or average) value about which the bell curve is centered. Because the normal curve is a symmetrical probability function, the mean value (which defines the median in terms of probability) is the same as the average value of the measured quantity. The average is simply the sum of all values divided by the number of values.

The standard deviation or variance (σ) represents the spread of the data about the average. Calculating the standard deviation is more complex:

$$\sigma = \sqrt{\frac{\Sigma x^2 - n(\mu)^2}{n-1}}$$

where σ is the standard deviation, x is the value of each sample, Σx^2 is the sum of the squared values for all of the samples, n is the total number of values, and μ is the average value.

Computer spreadsheet programs can simplify the calculations of the average and standard deviation from data. For example, Excel has built-in functions of "AVERAGE" and "STDEV" to calculate the average and standard deviation, respectively, for data analysis with a normal distribution.

The normal curve distributes the probability of any outcomes from a measurement or sample. The population of values can range continuously from $-\infty$ to $+\infty$, but the most probable values are centered about the mean. The total area under the curve is also equal to 1, because the probability of some value occurring (anywhere from $-\infty$ to $+\infty$) is 100%, or 1. This has important implications for how the normal curve can be applied.

Up to now, the focus has been on using statistics to describe observations (either during a test, by collecting inspection data, or by examining other quantitative data). Now consider techniques to use statistical data to predict future outcomes. The normal curve represents a population, and from it, one can determine the likelihood that a particular sample will be above or below a value on the x-axis. This determination is based on how many standard deviations a value is from the mean.

A characteristic of the normal curve is that as one moves farther away from the mean in either direction, the likelihood of attaining that value becomes smaller. Distances from the mean in standard deviations can be measured to express this concept. For example, approximately 68% of the area beneath the normal curve lies within one standard deviation on either side of the mean (Fig. 16.4). This indicates that the probability of being within one standard deviation from the mean is approximately 68%. It also indicates that the probability of a value being more than one standard deviation away from the mean (on either side of the mean) is only 32% (or 100% minus 68%, because the example involves 100% of the area beneath the curve). To know the probability of a value being more than one standard deviation below the mean, look at half of the distribution and see that the probability is half of 32%, or 16%. The same procedure can be done for two standard deviations. The probability of being more than two standard deviations away from the mean (on either side of the mean) is only approximately 5%. The probability of being more than three standard deviations away from the mean (on either side of the mean) is only approximately 0.3%.

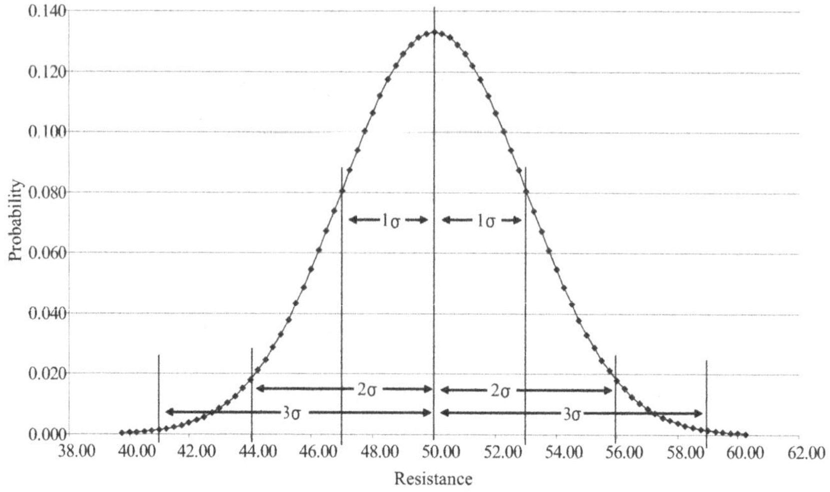

Fig. 16.4 Using normal curves to predict probabilities. The standard deviation concept can be used to predict the likelihood of values being greater or less than selected points on the normal curve. This curve has a mean of 50 Ω and a standard deviation of 3 Ω.

Expressing the distance from the mean in standard deviations allows the use of the normal curve probability distribution characteristics to determine the probability of being above or below any particular value.

In the failure analysis context, the normal distribution and its related probability distribution function are extremely important. Suppose the failure analysis team collects the range of outputs provided by a system for a parameter of interest and calculates the mean and standard deviation, or it already has information on the mean and standard deviation of that parameter. The failure analysis team can then compare a suspect value to the other outputs (using their mean and standard deviation) to know how likely it is that the suspect value will occur.

In the Fig. 16.4 distribution, for example, the mean is 50 Ω, and the standard deviation is 3 Ω. Suppose that a sample resistor has a resistance of 53 Ω. This is only one standard deviation away from the mean. Visually examining Fig. 16.4 and using the approximate percentages described in the preceding paragraphs shows that at 53 Ω, approximately 84% of the values lie below 53 Ω, and approximately 16% lie above it. It seems reasonable to expect a value of 53 Ω to occur.

Computer spreadsheet programs also provide functions to calculate the likelihood of a value occurring on a normal distribution. In Microsoft Excel, for example, the equation for making this assessment is:

$$P_{DABM} = \text{MIN}(\text{NORMDIST}(\text{tv}, \mu, \sigma, \text{true}), 1 - \text{NORMDIST}(\text{tv}, \mu, \sigma, \text{true}))$$

where P_{DABM} is the probability of the test value occurring this distance away from the mean, MIN is the Excel function for finding the minimum of values in the parentheses, NORMDIST is the Excel function for the normal distribution, tv is the test value (i.e., the value being compared to the distribution), μ is the mean, σ is the standard deviation, and true is the input used to select the NORMDIST function cumulative probability function.

For example, if the failure analysis team had a resistor with a resistance of 55 Ω, the preceding formula shows that such a value would have approximately 5% of the values occurring above it (and 95% of the values occurring below it). Similarly, if the team had a resistor with a resistance of 48 Ω, approximately 25% of the values would occur below it (and 75% of the values would occur above it). In both situations, the failure analysis team could safely conclude that these values probably occurred as the result of randomness about the mean.

However, if the failure analysis team had a resistor with a resistance of 62 Ω, the preceding formula indicates that the percentage of values expected to occur this far away from the standard deviation is extremely small (approximately 0.0004). In that case, the team could safely conclude that the likelihood of such a value occurring due to randomness alone is extremely small. It is probably not from the same parent population, or something else made the resistor move so far away from the mean.

Reliability and Probability of Passing

Reliability refers to the probability that a system will perform its intended function. As explained earlier, probability can range from 0 to 1, with 0 meaning something will never happen, and 1 meaning it will never fail. A probability of 1 is usually referred to as 100%. If a product has a reliability of 0.95, it means it is 95% likely to work. Stated differently, it means it has a 5% chance of failing.

Many products, especially electronics or weapons systems procured by the government, have specified reliability requirements. A typical requirement on a munitions item, for example, may be that it must be 95% reliable. During engineering development, the companies providing such products often prepare reliability analyses to show the predicted reliability of the product. These analyses are based on component failure rates and other factors described earlier. In many cases, companies also test products during engineering development to show compliance with specification reliability requirements.

Problems can emerge when products make the transition from engineering development to production if the production lot-acceptance test requirements are inconsistent with the product reliability. This frequently occurs when the production tests are based on attributes data instead of variables data.

Attributes data refer to a product either passing or failing rather than having a numerical value that can be analyzed using normal distribution characteristics. Attributes may be expressed as pass/fail, go/no go, accept/reject, and so on. Variables data collect quantitative information about product performance, which can be analyzed to determine means and standard deviations. Variables data allow determination of the likelihood of a product exceeding or falling below its acceptance criteria. It is much harder to do this with attributes data.

Early work by Dodge and Romig associated with sampling plans resulted in a military standard (MIL-STD-105, *Sampling Procedures and Tables for Inspection by Attributes*) used for determining sample sizes based on the production lot size and other factors. ANSI/ASQC Z1.4-1993, *Sampling Procedures and Tables for Inspection by Attributes,* replaced MIL-STD-105. The military recognized inherent deficiencies with both standards and the attributes-based sampling these standards represented, so the Department of Defense replaced MIL-STD-105 with MIL-STD-1916, *DoD Preferred Methods for Acceptance of Product.* This latter specification relies more on the statistical approaches outlined earlier in this chapter. That notwithstanding, the military still relies on sampling approaches with a specified pass/fail criteria for attributes inspections, as do many other nonmilitary organizations.

The problem emerges when selecting or defining a sampling plan. It is intuitive to believe that one ought to be able to test a reliable product without having any failures. A common approach is to select a sampling plan of 32 samples (based on tables from MIL-STD-105 or its derivative standards), with the acceptance criteria being accept on 0 failures and reject on 1 failure.* Intuitively, it feels that if a product is reliable, one ought to be able to pull a sample of 32 and experience no failures, thereby demonstrating that the lot is good. The intent is good, but the math is not.

The binomial distribution defines the probability of passing a test such as the one described here. It is:

$$\text{Probability of passing} = ((N!)/(n! * (N-n)!)) * p^{n} (1-p)^{(N-n)}$$

where N is the number of trials, n is the number of successes, and p is the product reliability.

The preceding calculation is relatively straightforward, but it can be simplified even further by using the Excel BINOMDIST function that greatly speeds the probability of passing calculations. When using the BINOMDIST function in Excel for these calculations, the expression and its argument are:

= BINOMDIST (number of samples required to pass, total number of samples, product reliability, false)

*The selection of such sampling requirements is typically based on lot size. Production lots of a few hundred to a few thousand components fall into the range that mandates a sample size of 32, with the reject criteria being 0/1 (as described earlier).

The false value tells the program that the operator wants to know the probability of exactly the number of samples passing that is specified.

In the preceding sampling situation, for a product with a reliability of 95%, the BINOMDIST function can be used as follows:

Probability of passing = BINOMDIST(32, 32, 0.95, false) = 0.194

The probability of passing a test with 32 samples with no failures, with a product reliability of 95%, is only 19.4%. This is not very high. It means that if the product is 95% reliable and is tested as described previously, it would fail the test approximately four out of five times.

Figure 16.5 shows a graph prepared using the Excel BINOMDIST function with the preceding sample size and accept/reject criteria to obtain a feel for how the probability of passing varies with product reliability. As seen in Fig. 16.5, the probability of passing a test such as the one outlined drops precipitously as the product reliability decreases.

In the situation described previously, the product lot-acceptance test requirements are inconsistent with its reliability. There are only four approaches for such situations:

- The manufacturer can continue to produce and test the product as outlined previously, but this quickly becomes expensive. More than 80% of the production lots would fail in such a situation.
- The manufacturer can attempt to convince the customer to buy the product when it fails the test. Although it sounds incredible, this actually occurs sometimes when the customer needs the product.

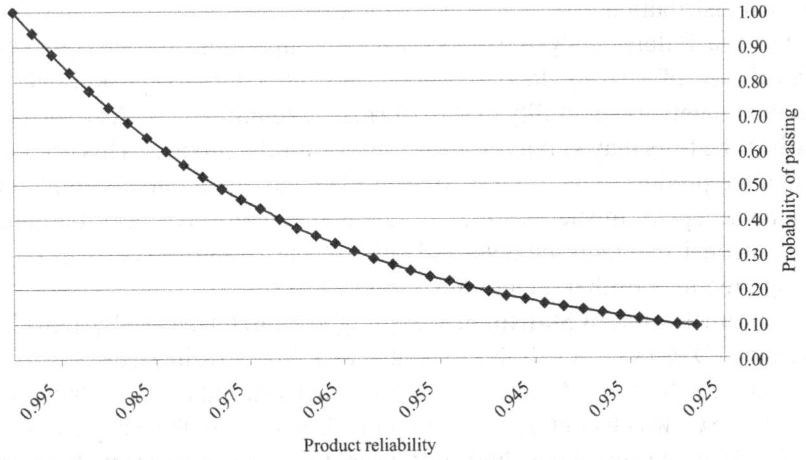

Fig. 16.5 Probability of passing versus product reliability. To have zero failures in a test with 32 samples, the product reliability must be extremely high. As the reliability decreases below 100%, the probability of passing falls quickly.

- The manufacturer can redesign the product to make it more reliable. This is also expensive, and, as seen in the Fig. 16.5 graph, the product reliability would have to increase from 95% to something well over 99.5% to pass the test most of the time.
- The customer can relax the test requirements to make the pass/fail criteria less onerous or adopt a variables data approach to obtain a better feel for the product acceptability.

Series Tests. Most production lot-acceptance tests address more than one performance characteristic, and the product must pass all of its tests for the manufacturer to sell the production lot. From a probability perspective, the probability of passing all such tests is equal to the product of the probabilities of passing each of the tests. For example, if a product must pass four performance characteristic tests, the probability of passing all four tests is equal to:

$$\text{Probability of passing} = P_1{*}P_2{*}P_3{*}P_4$$

where P_1 is the probability of passing the first performance characteristic test, P_2 is the probability of passing the second performance characteristic test, P_3 is the probability of passing the third performance characteristic test, and P_4 is the probability of passing the fourth performance characteristic test.

For variables-based test approaches, the probability of passing each test can be determined using the characteristic mean and standard deviation (and comparing these to the acceptable performance bounds, as outlined earlier in this chapter). For attributes-based test approaches, the probability of passing each test can be determined with the previously outlined binomial distribution.

From a failure analysis perspective, it is important to determine the probability of passing lot-acceptance and other tests, especially if the product meets its reliability and performance requirements. As explained earlier, the tests may simply be too stringent for the product. This does not mean the product should not be improved or that the failures should simply be accepted. In such a situation, the manufacturer has been burdened with inconsistent requirements, and either the test approach or the product design requires further attention.

Demonstration of Statistical Sampling in Court. Defense System Corporation (DSC) won a U.S. Navy contract for Mk 216 ship-launched chaff dispensers. The Navy designed the chaff dispensers and provided the drawings to DSC, which contracted to build the dispensers to the Navy's design.

The Navy designed the chaff dispensers to be launched from shipborne guns. The dispensers burst at preselected heights and released an aluminum foil cloud to confuse incoming enemy missiles and aircraft.

The Navy's specification required the chaff dispensers to be 95% reliable. The Navy's chaff dispenser lot-acceptance test required selecting 32 chaff dispensers out of each production lot, firing them, and confirming that all 32 operated properly. If a single chaff dispenser failed, the production lot would be rejected.

DSC could not pass the chaff dispenser lot-acceptance tests after repeated attempts. The company confirmed that the chaff dispensers it built conformed to the Navy's drawings. DSC outlined the binomial distribution to the Navy, explaining that the chaff dispensers met their 95% reliability requirement, but the probability of passing the lot-acceptance test with a 95% reliable product was less than 20%. The Navy did not agree and refused to compensate DSC. After approximately 18 months, DSC ceased operations and discharged all of its production employees.

DSC's owners sued the U.S. Navy in federal court to recover its damages. During a lengthy trial at the U.S. Armed Forces Board of Contract Appeals, DSC presented its case to a federal judge. DSC hired an expert witness to explain to the court the binomial distribution and the low probability of passing an attributes test that required zero failures in 32 tests with a 95% reliable product.

Recognizing the difficulty most people have in grasping statistical concepts, the expert explained to the presiding judge that he could demonstrate the concept with a bag of beans. The bag contained 100 beans, with 95 dark beans representing good product and 5 white beans representing defective product. The expert explained that the likelihood of pulling 32 beans out of a bag with none of them being white was low. He offered to pull out 32 beans to demonstrate the concept.

The judge smiled and directed the Navy attorney to pull the beans from the bag held by the expert. The 17th bean that the Navy attorney removed from the bag was white. At the conclusion of the trial, the judge awarded $25.4 million to DSC's owners.

Summary

This chapter developed the concepts of statistics and probability, the normal curve, and the mean and standard deviation. The normal curve shows the distribution of data about the mean, and the standard deviation represents the spread about the mean. Knowing the mean and standard deviation allows predictions about future outcomes to be made or the probability of an existing outcome to be defined.

When sampling is based on attributes data, the probability of passing a test is defined by the binomial distribution. Unless the product reliability is extremely high, the probability of passing a sampling test with zero failures is low. The probability of passing all of several individual tests is the product of the probabilities of passing each of the individual tests.

The failure analysis team can determine probabilities of passing tests using the Excel NORMDIST function (for variables data) or the BINOMDIST function (for attributes data).

The failure analysis team should assess the product reliability requirements and compare those to the test requirements. It may be that the test requirements are inconsistent with the product reliability requirements. In that case, the failure is an expected event. To preclude future failures, the product must become more reliable (which may not be feasible), or the test requirements must be relaxed to be consistent with the product reliability.

Example for Group Discussion. Why do you think the Navy refused to acknowledge the inconsistent reliability and lot-acceptance test requirements of the Mk 216 chaff dispenser? What could DSC have done differently to avoid finding itself in the position it did?

SELECTED REFERENCES

- G.A. Kimble, *How to Use (and Misuse) Statistics,* Prentice-Hall, Inc., 1978
- I. Miller and J.E. Freund, *Probability and Statistics for Engineers,* Arthur D. Little, Inc., 1965
- T.A. Payne, *Quantitative Techniques for Management,* Reston Publishing Company, Inc., 1982
- *Sampling Procedures and Tables for Inspection by Attributes,* ANSI/ASQC Z1.4-1993, ASQ, 1993

CHAPTER 17

Design of Experiments

AFTER THE FAILURE ANALYSIS TEAM hypothesizes failure causes, prepares a failure mode assessment and assignment (FMA&A), and evaluates all potential failure causes, it may find in some cases that several causes are still suspect but cannot be confirmed. In this situation, an experiment is necessary to confirm or rule out suspected causes. The predominant methods for doing this are analysis of variance (ANOVA) and Taguchi methods (a more powerful technique based on ANOVA).

Analysis of Variance

ANOVA is an analytical technique that determines if differences in average performance are due to randomness alone or to randomness plus one or more special causes. The phrase "randomness or randomness plus some special causes" is an important distinction, because randomness is always present, even if a difference in average performance is due to one or more special causes. The concept is that the failure analysis team can compare two or more groups of product (one with the suspect cause present, the other without) using the ANOVA technique to determine if the cause makes a difference. ANOVA is a statistical method that evaluates variability using mathematical calculations to determine if the differences are due to one or more special causes rather than randomness alone.

The ANOVA technique consists of several steps:

- The failure analysis team defines the groups to be compared. As mentioned previously, one group should have the suspect cause present; the other should not.
- The failure analysis team defines the output variable to be compared. ANOVA requires a quantitative result as a basis of comparison. The output parameter is what ANOVA uses when it performs the calculations described subsequently.
- The failure analysis team calculates the mean of each group.
- The failure analysis team calculates the mean of the entire group.

- The failure analysis team calculates the column sum of the squares by finding the difference between the mean of each group and the mean of the entire group, squaring the result, multiplying the result by the number of data points in each group, and then adding the values. The column sum of the squares represents the variability of the group means from the overall mean.*
- The failure analysis team calculates the error** sum of squares by finding the difference between each data point and its group mean, squaring these values, and then adding them. The error sum of squares represents the total variability of individual values from their respective group means, and it is a reference point for the randomness due to normal statistical variation.
- The failure analysis team calculates the total sum of squares by finding the difference between each data point and the overall mean, squaring each difference, and adding the squared differences. The total sum of squares represents the total variability of individual data points from the grand mean.
- The failure analysis team defines the total degrees of freedom, which are the number of independent comparisons that can be made between all data points (Fig. 17.1). If there are two groups of data, each with 15 data points, there are a total of 30 data points, and 29 independent comparisons can be made.*** Thus, there are 29 degrees of freedom.
- The failure analysis team defines the factor degrees of freedom, which represent the number of independent comparisons that can be made between the groups (or columns) being analyzed. The groups are compared to assess the effects of a factor, so making a comparison between groups is making a comparison of the effect of the factor. There are two groups, so there is one independent comparison and thus one factor degree of freedom.
- The failure analysis team calculates the error degrees of freedom. The total degrees of freedom equal the factor degrees of freedom plus the error degrees of freedom. In this example, there are 29 total degrees of

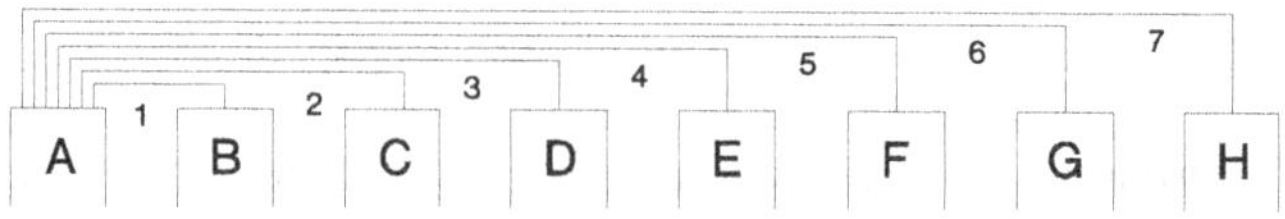

Fig. 17.1 The term *degrees of freedom* refers to the number of independent comparisons. This illustration has eight data points with seven degrees of freedom, because any other comparison between any two data points is already defined by the other comparisons; it would not be independent.

---

*A column is the same as one of the groups to be compared.

**Error is a statistical term for randomness.

***Any other comparison between any two data points is already defined by the other comparisons; it would not be independent.

freedom and 1 factor degree of freedom; therefore, there must be 28 error degrees of freedom.

- The failure analysis team calculates the column mean squares (MS), which are equal to the sum of the squares of the columns divided by the factor degrees of freedom.
- The failure analysis team calculates the error MS, which is equal to the sum of the squares for the error divided by the error degrees of freedom.
- The failure analysis team calculates the F-ratio, which is the column MS divided by the error MS.
- The failure analysis team finds a source for the F-ratio corresponding to the factor degrees of freedom and the error degrees of freedom. F-ratio sources can be found in statistics and other textbooks, on the internet, or by using Excel.
- The failure analysis team compares the calculated F-ratio to the source F-ratio and determines which is larger. If the calculated F-ratio is larger than the source F-ratio, the difference is said to be significant, and it is assumed to be due to the factor.* If the calculated F-ratio is smaller than the source F-ratio, the difference is said to be insignificant, and it is assumed to be due to randomness alone.

As the reader can infer from the preceding process, performing these steps is cumbersome and time-consuming. Excel makes these calculations much easier, and its data analysis pack can be used to greatly accelerate ANOVA.

An Excel ANOVA Example. Suppose circuit card assemblies are built with resistors from two different resistor lots. The circuit card assemblies provide an output measured in milliamperes. The concern is that there is a difference in the output of the two circuit card assemblies. The goal is to determine if the difference is due to a special cause (in this case, the two different resistor lots) or to randomness alone. After segregating two groups of 15 circuit card assemblies, the data are entered in an Excel spreadsheet, as shown in Fig. 17.2.

After entering the data into the spreadsheet, the Excel user clicks on "Tools," "Data Analysis," and "ANOVA Single Factor" to obtain the drop-down menu shown in Fig. 17.3.

After the "OK" button is clicked in the Fig. 17.3 menu, the spreadsheet creates and displays the table shown in Table 17.1.

There is some risk of reaching an erroneous conclusion based on the preceding process. In the example discussed here, the difference between the two circuit card groups is significant. The assumption is that the difference is due to the different resistor lots. There could be other variabilities between the circuit card groups that induce the difference. For this reason,

*Actually, the difference between the average performance of the two groups in such a case would be due to randomness plus the effects of the factor. For the sake of simplicity, it is said to be due to the factor, because the ANOVA is performed to determine if the factor makes a difference.

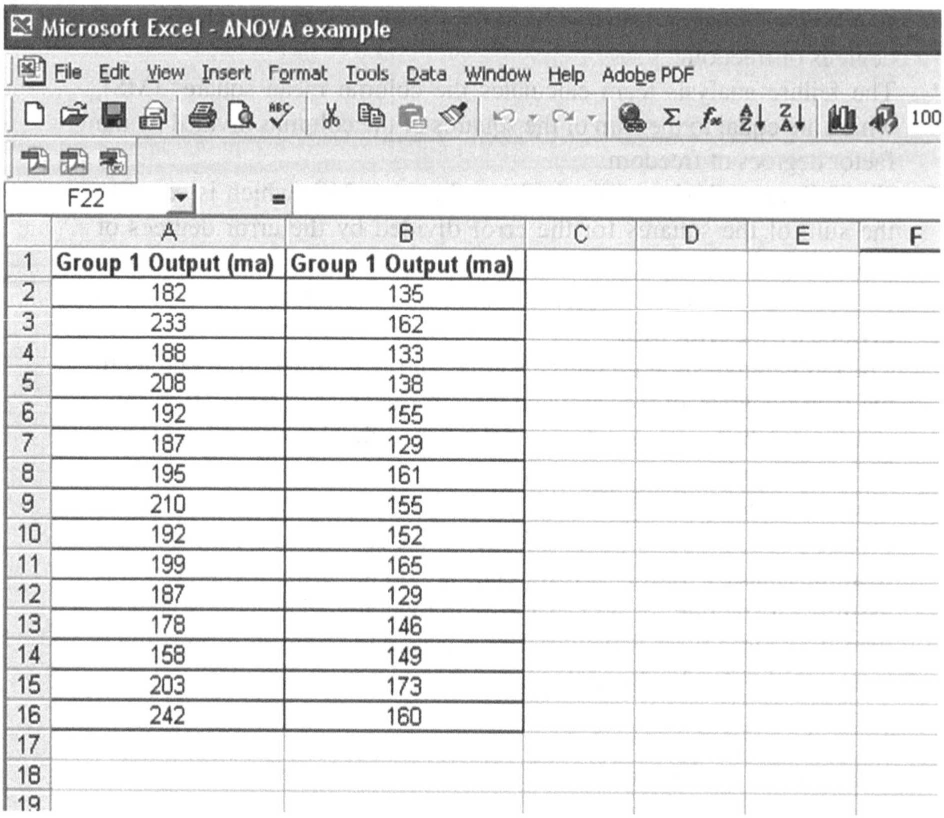

	A	B
1	Group 1 Output (ma)	Group 1 Output (ma)
2	182	135
3	233	162
4	188	133
5	208	138
6	192	155
7	187	129
8	195	161
9	210	155
10	192	152
11	199	165
12	187	129
13	178	146
14	158	149
15	203	173
16	242	160

Fig. 17.2 Circuit card assembly output data. The spreadsheet shows data entered for two groups of circuit card assemblies. The two circuit card groups used different resistor lots.

when the failure analysis team uses ANOVA, it should make sure that the only difference between the groups is the factor being evaluated.

Taguchi Testing

The preceding text discussed using ANOVA to determine if an average performance difference was due to randomness alone or a special cause. If the average performance difference is significant, it is desirable to determine if it is due to a single factor or if multiple factors are at work (either by themselves or as the result of interactions between the factors). Taguchi techniques allow the evaluation of several factors (and interactions between factors) simultaneously while minimizing the sample sizes and test runs.

When the failure analysis team converges on several remaining suspected causes, it can design an experiment using Taguchi techniques to determine which of the causes is significant. Designing a Taguchi experiment is a nine-step process, as shown in Fig. 17.4.

The first step in designing a Taguchi experiment involves selecting factors and interactions or the parameters to be evaluated during the experiment.

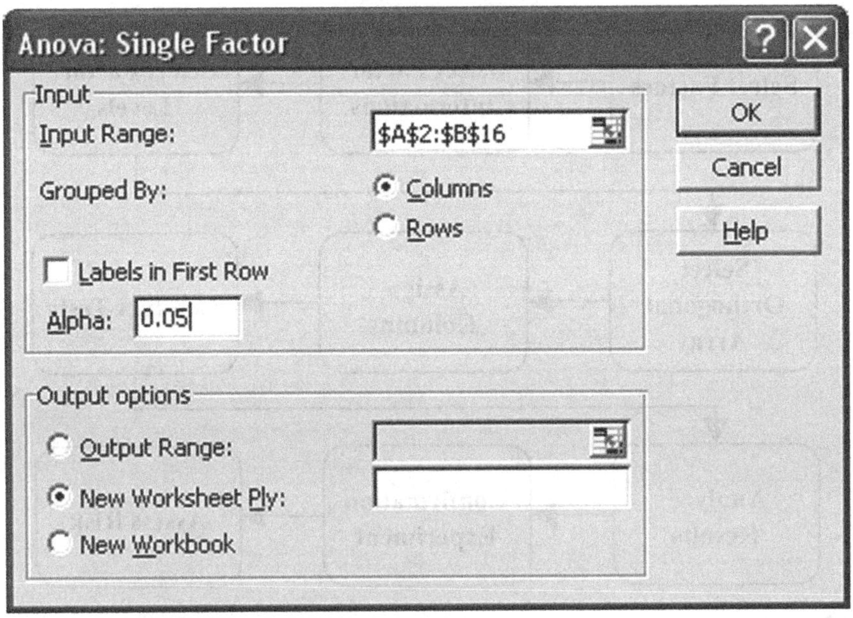

Fig. 17.3 Excel analysis of variance (ANOVA) menu. The input range highlights the cells containing the data. The user simply clicks and drags across the cells containing the data, and this block fills in automatically. "Alpha" defaults to 0.05, and it represents the confidence level at which the ANOVA is performed. (This means there is a 5% chance the conclusions are incorrect.) The user can lower this number if a higher confidence level is desired.

Table 17.1 Analysis of variance (ANOVA) table calculated using the Excel ANOVA data analysis tool

"F crit" is the value calculated by Excel for comparison. (It is the value found in a table of F-ratios based on the degrees of freedom.) "F" is the value calculated for the two groups of circuit card assemblies. Because "F" is greater than "F crit," it may be concluded that the output difference is due to a special cause.

ANOVA: Single factor
Summary

Groups	*Count*	*Sum*	*Average*	*Variance*
Column 1	15	2954	196.9333333	435.2095238
Column 2	15	2242	149.4666667	194.6952381

ANOVA

Source of variation	*SS*	*df*	*MS*	*F*	*P-value*	*F crit*
Between groups	16898.13333	1	16898.13333	53.65297853	5.63645E-08	4.195982228
Within groups	8818.666667	28	314.952381			
Total	25716.8	29				

These are usually the potential failure causes that cannot be either ruled in or ruled out using the methods outlined in the preceding chapters.

Selecting factor levels involves determining the values at which to set the factors during the experiment. It is recommended that several factors are selected during the initial Taguchi experiment, and each factor is set at two levels. This identifies which factors make a difference, and subsequent

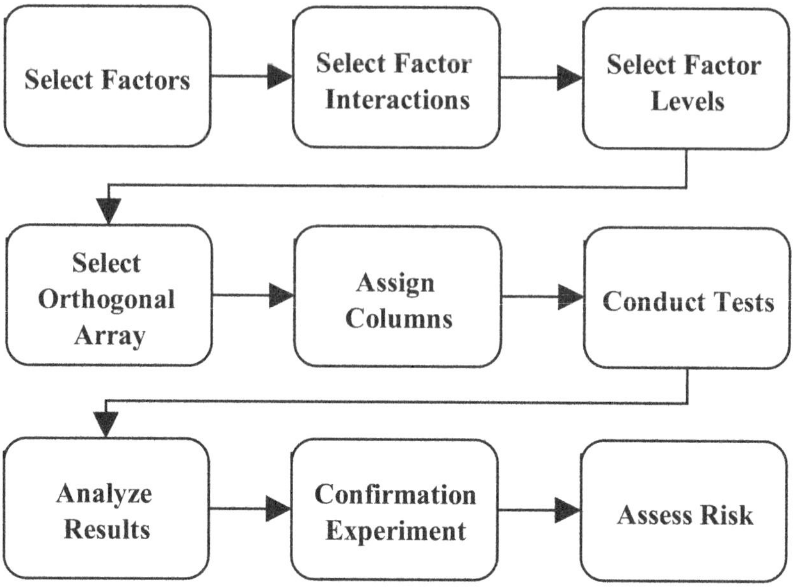

Fig. 17.4 The Taguchi design of experiments process selects factors and their interactions, develops an appropriate experimental design, analyzes results using the analysis of variance technique, and concludes with a confirmation experiment.

experiments (if necessary) establish exact design levels. The failure analysis team should consult development engineers and use best- and worst-case estimates to establish these levels.

Taguchi made it easy to select the appropriate orthogonal array. Taguchi texts have several orthogonal arrays, and selection of the appropriate array is based on the number of factors, interactions, and levels for each factor. For most failure analysis experiments, the L8 orthogonal array works well. The L8 orthogonal array is the one presented in the example included in this chapter.

After selecting the appropriate orthogonal array, the next step is to assign factors and factor interactions to each column. The following example shows how this is done.

The predefined Taguchi orthogonal arrays are one of the features that make the technique easy to use. The Taguchi orthogonal arrays have "1" and "2" values assigned in each cell, representing two levels for each factor. After the factors have been assigned to the columns, the "1" and "2" values in the Taguchi orthogonal array determine the test specimen configurations.

Another critical step in performing a Taguchi test is defining what will be measured as a quantifiable output variable. The output variable is used to measure the effects of the various factors. If the failure analysis team chooses the wrong output variable, the results of the experiment will be meaningless. The output variable should be something that can be tied to the success of the device being tested. The failure analysis team can instrument several output responses and perform an ANOVA on each to mitigate risk in this area.

After the failure analysis team has identified the factors, selected the appropriate orthogonal array, determined the factor levels, and identified the output parameter, the team can prepare the test specimens, perform the test, and analyze the results.

After completing the aforementioned, Taguchi recommends repeating the experiment. This is intended to minimize the risk of erroneous results. In the author's experience, however, there has not been a confirmation experiment with different results from that of the original experiment.

The Navy Guillotine Taguchi Experiment. The U.S. Navy uses an aerial refueling system based on the probe-and-drogue concept. This approach involves extending a hose with a metal basket on the end (the drogue). The aircraft to be refueled flies to the drogue and inserts a probe into it. When the connection is established, fuel passes from one aircraft to the other. After completing the transfer, the aircraft that was refueled breaks the connection. The tanker aircraft then rewinds the hose.

In the event of a failure that prevents rewinding the hose, the aerial refueling system includes a guillotine. The guillotine (located at the end of the aerial refueling system) surrounds the hose. Its purpose is to cut, clamp, and release the hose if the tanker aircraft pilot so commands. The reason is that if the hose fails to rewind, the guillotine mechanism allows the tanker aircraft to land without a fuel-laden hose trailing behind.

The guillotine operates by firing two electrically primed gas generators. When the gas generators fire, they create pressure that drives a piston connected to a blade and a clamp. When the blade and clamp encounter the hose, the blade cuts it, and the clamp seals the hose end that remains with the aircraft. Figure 17.5 shows the guillotine assembly.

The guillotine had been in production for more than 10 years; during that time, it experienced numerous failures. Sometimes, the guillotine cut the hose but failed to clamp it (and fuel leaked out). Other times, the cut was incomplete. The guillotine manufacturer attempted numerous design modifications related to the clamp and the blade, but none of these modifications fixed the intermittent failures to cut and clamp the hose.

The aerial refueling system manufacturer assigned a failure analysis team to investigate and correct the guillotine problems. The failure analysis team prepared a fault-tree analysis with the top undesired event, "Guillotine does not cut and clamp." After preparing the fault-tree analysis, the failure analysis team discovered that there were over 150 potential failure causes. After preparing an FMA&A, the team ruled out all but three of the potential failure causes:

- *Cartridge ignition simultaneity:* The team knew that two cartridges generated pressure to move the piston. They suspected that failure of the cartridges to fire simultaneously would degrade performance. The team

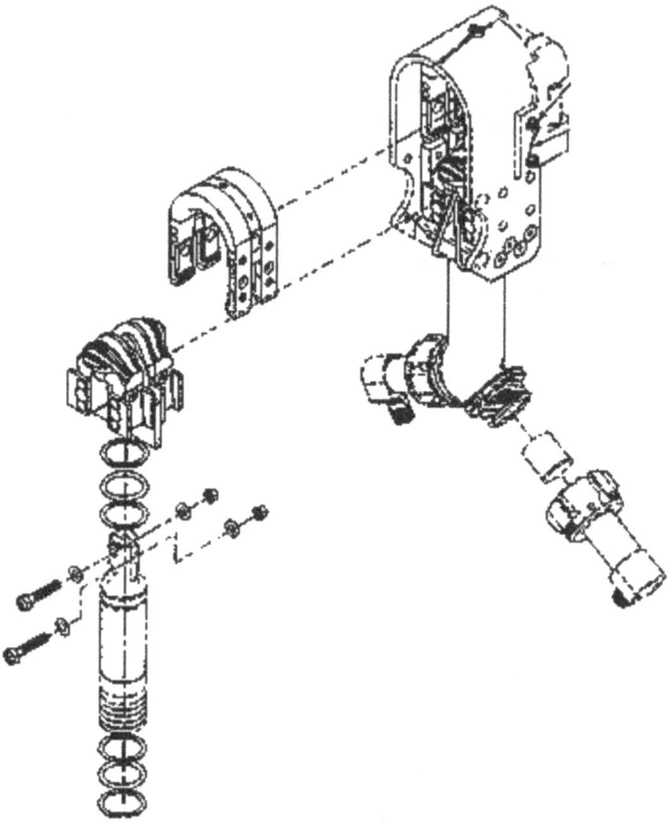

Fig. 17.5 Aerial refueling system guillotine. The guillotine surrounds the hose at the end of the aerial refueling system. Its job is to cut and clamp the hose during an emergency hose jettison.

recognized the need to determine if nonsimultaneous firing had an effect on guillotine function.

- *Cartridge lot:* Lot-to-lot cartridge output variability was also a potential cause of the guillotine failures. The failure analysis team knew that such variability existed, but they did not know if there was enough to influence guillotine function.
- *Ignition current level:* The guillotine cartridges are electrically fired, and one of the fault-tree potential failure causes included low current inputs into the cartridges. The team recognized a need to evaluate the effects of varying ignition current levels on guillotine performance.

The failure analysis team did not suspect interactions between any of the aforementioned factors, so it did not include this Taguchi capability in the experiment.

Based on the selection of the preceding three factors, the failure analysis team prepared the Table 17.2 test matrix using the Taguchi L8 orthogonal array.

Table 17.2 Taguchi L8 orthogonal array

The orthogonal array columns have been assigned for evaluating three hypothesized guillotine failure causes. Each column represents a different factor. The unused columns can be used for evaluating factor interactions. The "1" and "2" values in the array interior represent different factor levels.

	Column							
Trial	A Ignition simultaneity	B Cartridge lot	C Not used	D Ignition current	E Not used	F Not used	G Not used	Data Peak pressure
1	1	1	1	1	1	1	1	7140
2	1	1	1	2	2	2	2	7560
3	1	2	2	1	1	2	2	7980
4	1	2	2	2	2	1	1	7560
5	2	1	2	1	2	1	2	7000
6	2	1	2	2	1	2	1	7280
7	2	2	1	1	2	2	1	7700
8	2	2	1	2	1	1	2	7490

Next, the failure analysis team needed to determine factor levels. Their rationale is summarized as follows:

- The team felt that ignition either was or was not simultaneous. They specified these two conditions for the experiment. Simultaneous ignition was denoted as level 1, and nonsimultaneous ignition was denoted as level 2.
- For the cartridge lot factor, the team selected an older cartridge lot and a new cartridge lot. New lot cartridges were denoted as level 1, and older cartridges were denoted as level 2.
- For the current-level factor, the failure analysis team found that the lowest level at which the cartridge fired was 3 A, and the normal operating level was 10 A. The team selected 3 A for level 2 and 10 A for level 1.

When the factors and factor levels are determined, the Taguchi matrix determines the test configuration. The Table 17.2 orthogonal array for the first test specimen shows that cartridge ignition simultaneity is at level 1 (simultaneous ignition), cartridge lot is at level 1 (new cartridges), and ignition current level is at level 1 (10 A). The same can be done for each of the following seven test specimens to similarly define their configurations.

The failure analysis team selected peak pressure as the output parameter. The team needed a factor related to the device functioning successfully, and this factor must be something that could be quantified. Simply examining the guillotines to determine if they cut or clamped was not adequate, because ANOVA could not be performed on these attributes characteristics.

After collecting the output data, the failure analysis team prepared the ANOVA. A Taguchi experiment uses a simplified formula for the sums of squares for each of the columns:

$$SS_x = (\Sigma X_1 - \Sigma X_2)^2/N$$

where SS_x is the sum of the squares for each column, ΣX_1 is the sum of the values in the column at level 1, ΣX_2 is the sum of the values in the column at level 2, and N is the total number of specimens (eight in the guillotine case).

The ignition simultaneity sum of the squares is calculated as:

$$\Sigma X_1 = 7140 + 7560 + 7980 + 7560 = 30{,}240$$

$$\Sigma X_2 = 7000 + 7280 + 7700 + 7490 = 29{,}470$$

$$SS_A = (\Sigma X_1 - \Sigma X_2)^2/N = (30{,}240 - 29{,}470)^2/8 = 74{,}112.5$$

The team similarly calculated values for the other factors to find that the sum of the squares for the old versus the new cartridge lots was 382,812.5, and the sum of the squares for ignition current levels was 612.5.

Based on the Taguchi experiment results, the failure analysis team concluded that cartridge lot differences were significant. Ignition simultaneity also played a role, although it was not as significant as the cartridge lot. Ignition current level did not make much of a difference.

Armed with this knowledge, the team recognized that the guillotine was sensitive to cartridge lot differences. Subsequent investigation revealed wide variability between the old and new lot output pressure, and the cartridge specification did not specify a minimum output. The team found that ignition nonsimultaneity could be induced in an experimental environment, but the production guillotines always fired the pressure cartridges simultaneously. Based on this finding, the team concluded that no ignition simultaneity corrective action was required. The team revised the cartridge specification to include a minimum pressure output requirement (similar to the output of the older cartridge lot), and the guillotine failures ended.

Summary

The failure analysis team frequently converges on suspected causes that cannot be ruled in or out based on an examination of the failed hardware or on other techniques covered earlier in this book. In such situations, the failure analysis team can perform an experiment to assess if suspected causes induced the failure.

ANOVA is an analysis technique that determines if a difference in average performance is due to randomness alone or to randomness plus one or more special causes. The concept is to test groups with and without the suspect factor present and use ANOVA to determine if the difference in average performance is significant. Manually preparing an ANOVA is cumbersome, but the Excel ANOVA capability greatly accelerates the process.

Taguchi testing (based on ANOVA concepts) can simultaneously evaluate several suspect factors and provide statistically meaningful results. The use of Taguchi techniques requires identifying the factors and factor levels, selecting the appropriate Taguchi orthogonal array, assigning factors to the columns, selecting a quantifiable output parameter, assembling the test specimens, testing, and using sums of the squares techniques to assess the results.

SELECTED REFERENCES

- T. Payne, *Quantitative Techniques for Management,* Reston Publishing, Inc., 1982
- P. Ross, *Taguchi Techniques for Quality Engineering,* McGraw-Hill, 1988

CHAPTER 18

Corrective Action

THE FOUR-STEP PROBLEM-SOLVING PROCESS (Fig. 18.1) was introduced at the beginning of this book. Up to this point, this book has been focused on the first two steps (defining the problem and defining the problem causes). After accomplishing the activities described in the preceding chapters, the failure analysis team most likely has defined the causes of the specific failure that triggered the failure analysis as well as many other potential causes. This chapter addresses the last two steps (developing potential solutions and selecting the best solution), which involve corrective action. Corrective action means the implementation or modification of things designed to eliminate or reduce failure causes. These can include:

- Procedures
- Processes
- Workmanship practices
- Inspections
- Tests
- Specifications
- Engineering drawings
- Materials
- Tooling
- Production equipment
- Training
- Suppliers

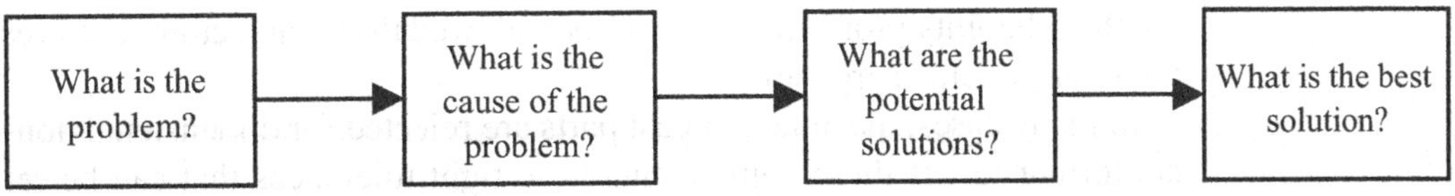

Fig. 18.1 The four-step problem-solving process. This chapter focuses on the last two steps (identifying possible solutions and selecting the best solution).

Corrective Action—Order of Preference

Some corrective actions are more preferable than others. The most desirable corrective actions are those that change the design or process such that the failure cause is eliminated. These do not rely on human beings not making mistakes. Examples include design changes that improve component strength, irreversible assembly features to preclude incorrect assembly, automated inspections that do not rely on human inspectors, and, where appropriate, requirements relaxation.

Less preferable corrective actions are those that rely on human beings doing things in precisely the right sequence without making any mistakes. These include such things as additional inspections, modifications to assembly procedures, or operating a product in a special manner to preclude failure.

In general, failure analysis teams should push for corrective actions that take human beings out of the loop, to eliminate the potential for human error. In practice, however, this is not always possible or economically viable. For example, if a manufacturer has a large inventory of potentially defective components and plans to implement a design change as a future corrective action, it may make sense to screen the existing inventory on an interim basis until the design change is incorporated. However, interim corrective actions of this nature should only be accomplished if the result of another failure is not catastrophic.

Design Changes. The failure analysis team may find that the root cause of the problem can be eliminated with a design change. An example is a bearing that failed because it was installed in a reversed manner. A design change that incorporates a taper on the bearing outer cone and a mating taper where the bearing is installed would prevent reversed assembly.

Requirements Relaxation. The failure analysis team may find that a failure involves not meeting a requirement, with the nonconforming condition being slightly out of the acceptable range. If extensive redesign or other extraordinary steps are required to correct the problem, it may make sense to relax the requirements, if the customer concurs.

In one example involving a military target designation laser, the customer required the laser to operate at a very high energy level. Operating the laser at these levels decreased its reliability. The manufacturer and the customer tested the laser at reduced energy levels and found that this allowed the system to meet all performance requirements while simultaneously eliminating the reliability problem. The customer allowed the manufacturer to lower the laser system energy levels.

In many cases, machined or cast parts are rejected for dimensional nonconformances. If these components have tight tolerances that can be relaxed without any next-assembly fit or functionality issues, it makes sense to do so.

Process Modifications. Sometimes, the manufacturing process causes failures. Chapter 4 described bulge-forming failures on aircraft fuel tanks. Stretching the material beyond the yield point induced these failures. The corrective action was a process modification that incorporated a two-step expansion operation, with an annealing operation in between, to prevent bursting. In effect, this was a design change, but to the process rather than the product. This is similar to a design change in that it does not rely on human beings doing things in an exact sequence, and, unless someone attempted to use the old process, it eliminated the failure cause.

Training. Sometimes, failures are caused by manufacturing technicians who assemble the product incorrectly. Sometimes, failures are caused by the person using the product incorrectly. In these cases, training may correct the problem. Chapter 14 described an optics debonding situation ultimately found to be caused by incorrect use of a solvent-based cleaning station. The manufacturing technicians did not know how to operate the cleaning equipment. Training the operators corrected the problem.

Training is important, and many failures are the result of inadequate training. However, training is not as good as a design or process change, because it relies on humans executing the process exactly as specified. Under stressful conditions, poor lighting, new employees, or any of a myriad of other human-error-inducing conditions, the human error that caused the failure may recur. Ideally, in the situation described previously, it is desirable to find a cleaning process that does not depend on daily solvent replacement.

Some manufacturing organizations take the training concept to a higher level by implementing certified operator programs, which involve defining the process, identifying all key performance characteristics, training the manufacturing technicians, testing the technicians to confirm they have mastered the process, and certifying them accordingly. This approach minimizes (but does not eliminate) the risk of human error.

Additional Tests and Inspections. This is a less desirable approach, because it relies on inspectors to find and cull out defective product. In some cases, as explained at the beginning of this chapter, such an approach may be necessary on an interim basis (as was done in the Chapter 10 dispenser fuze example). On a longer-term basis, however, relying on additional inspections and tests is not a good approach. Inspectors will not detect all of the defects.

Cautions or Warnings. A caution is a note or label indicating that damage to equipment can result if certain steps are not followed. A warning is a note or a label indicating that a human being can be injured or killed if certain steps are not followed.

Relying on cautions or warnings may be necessary in some instances, but again, these corrective actions rely on human beings not making mistakes. Whenever possible, the process or the design should be modified to eliminate the condition requiring the caution or warning.

Special Operations or Process Steps

Special actions can be used as corrective actions to overcome design, process, and other failure causes, but because these rely on freedom from human error, they are less desirable and should be used only as temporary or interim corrective actions.

When the U.S. Army introduced the M9 9 mm Beretta as a replacement handgun for the .45 automatic, it soon discovered that the initial pistols had a metallurgical defect in the slide. This could cause a broken slide, which presented a hazard to the shooter. The Army's interim corrective action was to limit the number of rounds fired by each M9 handgun. The longer-term solutions included replacing the defective slides and making changes to the process to prevent the metallurgical shortfalls.

Supplier Corrective Actions. Any of the corrective actions discussed previously can be considered for implementation at suppliers. In some cases, manufacturers simply return defective products to suppliers, and the suppliers provide replacement components or subassemblies. This may provide quick relief, but it is not a good practice, especially if the returns occur frequently. It is recommended that the manufacturer insist the supplier identify the failure cause and implement appropriate corrective actions.

Corrective Action Locations. Whenever the failure analysis team finds a failure cause and selects appropriate corrective action, the team should make sure that the corrective action is implemented in all affected areas. A common shortfall is implementation of corrective actions only on work in process or future work. When corrective action is identified, the failure analysis team should consider implementing it in the following locations:

- Supplier inventory
- Supplier work in process
- Supplier repair centers
- In-house inventory
- In-house work in process
- Planned future work
- In-house repair centers
- Other service centers
- Fielded systems

The Global View. The beginning of this chapter mentioned that the failure analysis team should select appropriate corrective action after identifying the failure cause. The team should take a more global view and recognize that a failure analysis exposes many more potential causes beyond those that caused the specific failure. The question becomes:

Does it make sense to implement corrective actions for potential causes that have not yet caused a failure?

In many cases, the answer to this question is "yes." At this point, the term *corrective action* could be replaced by the more correct term, *preventive action*. It may not make sense economically or from a risk perspective to implement preventive actions for every potential cause. The failure analysis team should make recommendations for preventive actions on causes it deems likely to occur or for those with unacceptable consequences. This is one of the great advantages of a systems failure analysis. The process finds and corrects the cause of a failure, and it also exposes and identifies actions to prevent other failures.

Corrective Action Implementation

The failure analysis team can use the failure mode assessment and assignment (FMA&A) for assigning corrective actions, just as it did for failure-cause evaluation actions. The FMA&A assignment column is a good location for listing required corrective actions and implementation dates and for tracking implementation status. It is also a good place to note required follow-up dates to ensure the corrective action has been implemented and that it has been effective. Sometimes, failure analysis teams identify appropriate corrective actions, but they are not implemented. Sometimes, the corrective actions are implemented, but the failures they were designed to correct continue (suggesting that either the failure-cause identification or the corrective action was incorrect). In either case, scheduling a follow-up date to determine if the corrective action has been implemented and if it is effective is a good idea.

Summary

Corrective action can include a change to procedures, processes, workmanship practices, inspection or testing approaches, specifications, engineering drawings, tools, production equipment, materials, or training to prevent or minimize failure causes. Ideally, corrective actions should not rely on freedom from human error, but sometimes, such actions are necessary as interim steps. Whenever possible, corrective actions should change the design, the process, or the requirements such that future failures cannot be induced by human error.

The failure analysis team must ensure that corrective actions reach all affected suppliers, inventory, work in process, items returned for repair, and fielded systems.

The failure analysis team should schedule a follow-up date to ensure corrective action implementation and that the corrective action has fixed the problem. The FMA&A is a good tool to use for this purpose.

Example for Group Discussion. Suppose your company develops and manufactures electrical harness assemblies, and you find that you are rejecting several each week for short circuits within the harness assembly.

After developing a fault-tree analysis and an FMA&A, you disassemble the harnesses and discover that the insulation on wires is nicked and torn. After interviewing the manufacturing technicians, you find that the harnesses have steel elbows that make 90° bends, and that the interior corners of these elbows are abrading the wire insulation. The damage occurs as the wires are pulled through the elbows during harness assembly. After checking the harness and elbow engineering drawings (the elbow is a procured item), you find there is no requirement to break the internal sharp edges of the elbows.

Your company has already built several hundred harnesses; several hundred more are in various stages of assembly. Your company has several thousand harness elbows in stock, all of which have been formally accepted. Your company has already paid for the elbows.

What corrective actions do you recommend?

SELECTED REFERENCES

- J. Berk and S. Berk, *Quality Management for the Technology Sector,* Newnes, 2000
- J. Ketola and K. Roberts, *Correct! Prevent! Improve! Driving Improvement Through Problem Solving and Preventive and Corrective Action,* ASQ, 2003
- D.E. Robitaille, *The Corrective Action Handbook,* Paton Press, 2001

CHAPTER **19**

Post-Failure-Analysis Activities

UP TO THIS POINT, the discussions on systems failure analysis have covered many topics, including:

- Defining the problem
- Unearthing potential failure causes using brainstorming, mind mapping, the "five whys" technique, Ishikawa diagrams, and fault-tree analysis
- Managing the failure analysis using failure mode assessment and assignment matrices
- Evaluating potential failure causes using pedigree analysis, change analysis, dimensional conformance and materials assessment tools, and analysis of variance and Taguchi test techniques
- Selecting and implementing appropriate corrective actions

At the conclusion of a systems failure analysis, the people involved should have a much more in-depth understanding of how the system is supposed to work. The analysis should help understand shortfalls in the design, production, testing, and use of the system. The team will have identified other potential failure causes and actions required to preclude future failures. This is valuable knowledge, and it should not be set aside or ignored when the failure analysis team concludes its activities.

Failure Analysis Libraries. All of the documentation resulting from each systems failure analysis should be organized into a failure analysis library. This library should be available as a reference tool to prevent future failures and to assist other failure analysis efforts on similar products. The library can be organized along product lines, processes, customers, or other factors significant to each business.

Design and Process Guidelines. Each failure analysis provided insights that can be used to improve products and processes and to prevent future failures. It is recommended that a "lessons learned" document be prepared to summarize these improvement opportunities. This document can be a resource on future product and process development efforts. The organization can also condense the results of its failure analyses into design guidelines. Most product development efforts are evolutionary (i.e., they are modifications and improvements to existing product lines). The use of previous failure analysis findings as guidelines for future development activities minimizes the risk of future failures. The organization can similarly prepare process guidelines based on previous failure analyses. This may have even greater application than design guidelines, because processes cut across product lines.

Troubleshooting and Repair Documents. Failure analysis findings, including all hypothesized failure causes, can be incorporated in troubleshooting and repair guidelines.

Failure Analysis Procedures. Many failure analysis procedures consist primarily of administrative guidelines for routing nonconformance documentation and other related forms, with little or no guidance on how to analyze failures. The approach outlined in this book can be used as the basis of a failure analysis process. The process should include the following steps:

- Designate a failure analysis team, with representatives from engineering, quality assurance, manufacturing, purchasing, and field service.
- Gather all related failure information.
- Review the aforementioned information, and define the problem.
- Identify all potential failure causes using brainstorming, mind mapping, Ishikawa diagrams, flow charting, the "five whys" technique, or fault-tree analysis.
- List each potential failure cause in a failure mode assessment and assignment.
- Use appropriate documentation reviews, interviews, design analyses, hardware analyses, and designed experiments to converge on the root cause.
- Identify potential interim and long-term corrective actions, and select the most appropriate corrective actions.
- Ensure corrective actions are implemented in all relevant areas (suppliers, inventory, work in process, repair centers, and fielded systems).
- Follow up after corrective actions have been implemented to assess corrective action efficacy.
- Evaluate other potential failure causes as corrective action candidates, and incorporate preventive actions where it makes sense to do so.
- Incorporate failure analysis findings into a failure analysis library, design and process guidelines, and troubleshooting and repair documents.

Finally, as noted in Chapter 1, failure analysis training should be incorporated in all technical and manufacturing organizations. In addition to becoming knowledgeable in failure analysis procedures, tools, and technologies, failure analysis training instills a way of thinking that helps engineers, manufacturing specialists, purchasing specialists, field service technicians, and quality assurance personnel become more successful.

APPENDIX **A**

Component Failure Causes and Suggested Analysis Methods

Table A.1

Suspected component failure cause	Suggested analysis method
Nonconforming dimensions	Dimensional inspection
Missing component	Visual examination, x-ray or n-ray (if device cannot be disassembled and depending on materials)
Incorrect material	Visual examination, along with energy-dispersive analysis of x-rays (EDAX), spectrometry, chromatography (depending on material type)
Incorrect assembly	Visual examination, x-ray or n-ray (if device cannot be disassembled and depending on materials)
Internal component presence and fit	Visual examination, x-ray or n-ray (if device cannot be disassembled and depending on materials)
Contamination	Visual examination, magnification, optical microscopy, scanning electron microscopy, along with EDAX, spectrometry, chromatography (depending on material type)
Interference	Dimensional inspection, tolerance analysis, visual examination, microscopic examination
Wear	Dimensional inspection, tolerance analysis, visual examination, microscopic examination, chemical and visual analysis of lubricant
Excessive clearance	Dimensional inspection, tolerance analysis, visual examination, microscopic examination
Deformation	Tolerance analysis, visual examination, dimensional inspection, stress analysis, strength testing
Porosity	Visual examination, microscopic examination, x-ray
Leaking	X-ray inspection for porosity, dimensional inspection, tolerance analysis for adequate seal/gasket squeeze, visual and microscopic examination for presence of contaminants or cuts on seals, materials analysis to confirm correct seal/gasket material, dye penetrant inspection, fluorescent dye pressure testing, magnafluxing, helium sniffer leak testing
Electrical/electronic component fails open/short	Visual examination, multimeter testing, replacement with known good component
Electrical connection fails open/short	Visual examination, multimeter testing, replacement with known good component
Electronic subassembly performs outside specified range	Visual examination, multimeter testing, subassembly testing, replacement with known good component
Electrical disconnect	Visual examination, magnification, optical microscopy
Fractures	Visual and microscopic examination, dye penetrant inspection, fluorescent dye pressure testing, x-ray
Mechanical disconnect	Visual examination, dimensional inspection, surface fractography, stress analysis, strength of materials testing

(*continued*)

Table A.1 (continued)

Suspected component failure cause	Suggested analysis method
Metallic component compressive, tensile, brittle, fatigue failure	Surface fractography, microscopic examination, hardness testing, materials analysis, stress analysis, dimensional inspection
Plastic or composite component compressive, tensile, brittle, fatigue failure	Surface fractography, microscopic examination, materials analysis, contaminant presence, stress analysis, dimensional inspection, storage history review
Ceramic component compressive, tensile, brittle, fatigue failure	Surface fractography, microscopic examination, hardness testing, materials analysis, stress analysis, dimensional inspection
Adhesive, epoxy, or composite disbond or delamination	Microscopic examination, Fourier transform infrared analysis to check for silicone or other contaminants, process review

Index

A

accelerated aging, 93
acid etching, 142
adhesive bonds
  application process for, 135
  bondline dimensions for, 119–120, 155
  cleanliness and, 120
  corrective actions for, 120
  cure time, 135
  curing time/temperature for, 120
  failure modes of, 119
  surface contamination and, 135
  surface-finish conditions and, 135
  tolerances of, 119–120
aerial refueling system, 177
aerojet, 87–92
aging, 102
  accelerated, 93
  changes from, 86–87
  crazing and, 123
  plastic component failures and, 123
air force, U.S., 21, 35–36, 57, 87
airborne contaminants, 142
analysis of variance (ANOVA), 171–174, 175(F), 180
AND gates, 39, 53, 55, 63, 66
  sequential, 50–51
ANSI, 166
Apache helicopter, 2, 154–156
Armed Forces Board of Contract Appeals, U.S., 169
army, U.S., 4, 155–156
arsenic, 2
attributes data, 165–166
Auger analysis, 102, 105, 141

B

basic failure event, 37
  probabilities for, 60–61
beach marks, 115
Bell Laboratories, 21, 35–36, 57
BINOMDIST, 166–167, 170
binomial distribution, 158–159, 166–167
Boeing, 21, 35–36, 57
bond-joint failures, 129
bond-joint leaks, 134–135, 138
bondline dimensions, 119–120, 155
brainstorming, 3, 25–27, 67
brittle fractures, 113–115
burning
  smell indicating, 124–125
  visual indicators of, 124–125
Buzan, Tony, 27

C

CAD. *See* Computer-aided design
calibration data, 77
capacitors, 125–126
cause-effect diagrams, 29. *See also* Ishikawa diagrams
cautions, 185
CBU-87/B cluster bomb, 87–92, 97
ceramic component failures, 124
certificates of conformance, 77–78
chaff dispensers, 168–169
Challenger (space shuttle), 1–2
change analysis, FMA&A and, 81
change sources
  aging, 86–87
  corrective actions, 184–185
  design changes, 82–83, 184
  environmental changes, 84–85
  lot changes, 85–86
  process changes, 83–84, 185
  supplier, 86–87
  test/inspection changes, 84, 185
charting symbols
  in fault-tree analysis, 37–39
  transfer symbols, 37, 39, 47–49

chemical analysis, 94
chemical concentration monitoring equipment, 142
chromatography, 94, 102, 105, 106(F), 141
circuit card assembly failure, 125–126
circuit equations, 151
circuit failure analysis, 151–152
circuit performance, 151–152
cleaning, ultrasonic, 143–145
cleaning solutions, 142
cleanliness, adhesive bonding and, 120
clearance, excessive, 149–150
cohesive failure, 120–121
command event, 37, 42–43, 47
  probabilities for, 61–63
  state-of-part v. state-of-system, 52–53
component failures
  ceramic components, 124
  composite components, 119–122
  corrosion, 110–111
  environmental factors in, 153–154
  fractures, 111–115, 153
  materials compatibility and, 153
  mechanical, 152–153
  metallic components, 110–119, 153
  plastic components, 123
composite structures, 155–156
composites component failures
  adhesive bonds in, 119–120
  adhesive/epoxy failure in, 120–121
  bondline dimensions in, 119–120
  cleanliness and, 120
  cohesive failure in, 120–121
  defined, 119
  prepreg materials and, 121–122
composition analysis, 94
compressive fractures, 115
computer modeling, stress analysis and, 101
computer-aided design (CAD), 150
conclusions, jumping to, 4
connection failures, 127–128
contamination, 129, 135
  airborne, 142
  corrosion and, 141
  detecting, 141
  of epoxy, 143
  failures caused by, 141
  human-borne, 142
  lubricant, 142–143
  sources of, 141–143
  surface, 135
  work area, 142
continuous statistical distributions
  exponential, 161
  log normal, 161
  normal, 160–165
  Weibull, 161
coordinate-measuring machines, 93, 98
corrective actions
  for adhesive bond failures, 120
  FMA&A and, 187
  for fractures, 111, 115
  global view of, 186–187
  human error and, 184
  implementing, 2, 3, 91, 187
  locations of, 186
  order of preference for, 184–185
  selecting, 2, 3
  special actions/process steps, 186
  supplier, 186
  types of, 183
corrosion
  bonded surface leaks and, 135
  circuit card assemblies and, 125
  contamination and, 141
  fretting and, 118
  gasket leaks and, 132, 134
  leaks and, 131
  materials compatibility and, 153
  process of, 110
  stress cracking and, 111
  structural leaks and, 138
  uniform, 110–111
  in valves, 137
cracking, 111, 119
crazing, 119, 123
curing, 120, 122, 135
cutting oils, 142

D

DC-10, 54
defects, as normal part of manufacturing process, 11
Defense System Corporation (DSC), 168–169
deformation
  of gaskets, 132
  of metal components, 115–117
  of plastic components, 123
degrees of freedom, 172
delamination, 119
Department of Defense, U.S., 61
design analysis
  circuit performance in, 151–152
  environmental factors, 153–154
  fatigue fractures and, 115
  materials compatibility in, 153
  mechanical failures and, 152–153
  requirements identification/verification in, 148–150
design changes
  as corrective action, 184
  failures due to, 82–83
design flaws, reasons for, 147, 156

design guidelines, 189–190
dimensional inspection, 93, 97–99
dimensions, 150
disbonds, 119, 143–144, 155
discrete statistical distributions
  binomial, 158–159, 166–167
  hypergeometric, 158
  Poisson, 159–160
disposition, 82–83
distribution loads, 111
documentation
  nonconformance, 76
  for product pedigree, 75–78
  from suppliers, 76–78
  troubleshooting/repair, 190
drawings, 77
  inspection criteria on, 97
  nonconformance with, 109
  requirements in, 148
ductile fracture, 113
dye penetrant inspection, 93, 99–100, 138

E

EDAX. *See* Energy-dispersive analysis of x-rays
80/20 rule. *See* Pareto principle
elastic limit, 115
electrical component failures
  capacitors, 125–126
  integrated circuits, 125
  resistors, 125
electrical connection failures, 127–128
electrical/electronic failures
  circuit card assemblies, 125–126
  connections, 127–128
  electrical components, 126–127
  electronic subassemblies, 124–125
electron microscope. *See* Scanning electron microscopy
electron probe microanalysis (EPMA), 102, 103
electronic subassembly failure, 124–125
energy-dispersive analysis of x-rays (EDAX), 94, 102, 103, 103(F), 141
environmental changes, 84–85
environmental conditions, 52–53
environmental factors, 153–154
environmental inspection approaches, 93, 100–102
EPMA. *See* Electron probe microanalysis
epoxy, 119
  application process for, 135
  contamination of, 143
  cure time, 135
  failure of, 120, 143–144, 155
  mixing problems, 134
  pot life of, 120, 135
  premixed v. mixed, 134
  shelf life of, 120, 134–135
  surface contamination and, 135
  surface-finish condition and, 135
error sum of squares, 172
etchants, 142
evaluating potential failure causes, 72
event probabilities, 57
events
  basic failure, 37, 60–61
  command, 37, 42–43, 47, 52–53, 61–63
  normal, 38, 60
  numbering, 44
  probabilities for, 57, 60–65
  secondary, 52–53
  state-of-part, 52–53
  state-of-system, 52–53
  undeveloped, 38, 42, 43, 61, 124
examination techniques
  categories of, 93–94
  chemical composition analysis, 94
  dimensional inspection, 93, 97–99
  environmental test approaches, 93, 100–102
  materials analysis, 94, 102–105, 106, 114, 132, 141
  mechanical inspection approaches, 93, 100–102
  nondestructive test approaches, 93, 99–100
  optical, 93–97, 106
Excel
  ANOVA in, 173–174, 175(F)
  AVERAGE function in, 163
  BINOMDIST function in, 166–167, 170
  NORMDIST function in, 164–165, 170
  STDEV function in, 163
Excel Visual Basic, 151
excessive clearance, 149–150
experiment design
  ANOVA, 171–174, 180
  Taguchi testing, 174–181
exponential distribution, 161

F

F-16 Air Combat Fighter, 23–24, 37, 53, 55, 65
F-35 fighter, 124
factor levels, 175–176, 179
failure analysis library, 3, 189
failure analysis process, 2–3, 190–191
failure analysis team, 2
  composition of, 8
  meeting frequency for, 72
  meeting length for, 26

failure causes
bond-joint leaks, 134–135
categories of, 29–30
contamination, 141
evaluating potential, 72
factor levels for, 175–176, 179
gasket leaks, 131–133
hypothesized, 69–70
identifying, 5, 6–7, 25
immediately adjacent, 40–42
o-ring leaks, 133–134
perspectives for, 109
point-of-failure, 39–40
ranking, 57
smell indicating, 124–125
in Taguchi experiment, 174–175, 177–178
failure event, basic, 37
failure mode assessment and assignment (FMA&A), 89, 177
assessment column, 70
assignment column, 70–71
change analysis and, 81
corrective action assignment with, 187
defined, 67
event, sheet column, 68–69
hypothesized failure cause column, 69–70
preparation for, 67–71
table for, 67–71
failure rate, 58, 60–61
fatigue failure, 101–102
corrosion and, 111
fractures and, 115
fatigue loading, 101
fatigue testing, 93
fault-tree analysis, 3, 21, 25, 67, 68(F), 89
charting symbols in, 37–39
command events in, 42–43
defining problem for, 39
development of, 35–36
event numbering in, 44
event probabilities in, 57
failure cause ranking in, 57
immediately adjacent causes, 40–42
INHIBIT gates in, 49–50
key points for, 36–37
point-of-failure causes, 39–40
probability determination in, 59–63
redundancy defeaters in, 53–55
sequential AND gates in, 50–51
state-of-part v. state-of-system events in, 52–53
systems failure analysis and quantification of, 65
top undesired event probabilities in, 64–65
transfer symbols in, 37, 39, 47–49
Fishbone diagrams. *See* Ishikawa diagrams
"Five whys" technique, 3, 25, 30–31, 67
flow charting, 3, 25, 31
fluorescent dye pressure testing, 93, 99–100, 138
fly-by-wire, 23, 53, 55
FMA&A. *See* Failure mode assessment and assignment
follow-up, 3
force-*versus*-deflection consistency, 154
fourier transform infrared spectroscopy (FTIR), 102, 103–104, 120, 141, 144
fractures
brittle, 113–115
of ceramic parts, 124
compressive, 115
corrective actions for, 111, 115
design analysis and, 115
ductile, 113
fatigue failure and, 115
identifying mode of, 111, 112(T)
materials analysis of, 114
materials compatibility and, 153
of metal parts, 111–115
stress-induced, 111
tensile, 113
F-ratio, 173
fretting, 118
FTIR. *See* Fourier transform infrared spectroscopy

G

gas generators, 177
gasket leaks, 131–133, 138
gates, 37
AND, 39, 50–51, 53, 55, 63, 66
INHIBIT, 39, 49–50, 52, 55, 61, 63, 66
OR, 38, 52, 55–56, 61–63
types of, 38–39, 42
general dynamics, 23–24, 37, 65
goodness-of-fit tests, 157
Guillotine mechanism, 177–180
Gulf War (1990), 1, 5

H

hardness testing, 93, 100–101, 106
Harley-Davidson, 138–139
hidden factory, 11
defining, 12
locating, 12–13
quantifying, 13–15
rejection rates and, 12–13, 15(T)
Honda, 139
human error, 38, 42, 52–53
corrective actions and, 184

probabilities for, 59–60, 66
training and, 185
human-borne contaminants, 142
humidity, 102
Hussein, Saddam, 1
hypergeometric distribution, 158
hypothesized failure cause, 69–70

I

ICBM. *See* Intercontinental ballistic missile
inadvertent system function, 21
INHIBIT gates, 39, 49–50, 52, 55, 63
probabilities for, 61, 66
inspection
categories of, 93–94
chemical composition analysis, 94
dimensional, 93, 97–99
dye penetrant, 93, 99–100, 138
environmental test approaches, 93, 100–102
equipment for, 98–99
magnetic particle, 99, 138
materials analysis, 94, 102–105, 106, 114, 132, 141
mechanical, 93, 100–102
neutron beam, 99–100
nondestructive test approaches, 93, 99–100
optical, 93–97, 106
pedigree analysis and, 76
records of, 76
inspection/test changes, 84, 185
integrated circuits (IC), failure of, 125
intercontinental ballistic missile (ICBM), 21, 35
intermittent system failures, 21
Ishikawa, Kaoru, 29
Ishikawa diagrams, 3, 25, 29–30, 67

K

kawasaki, 29

L

LANTIRN, 9, 11
laser target designators, 1, 184
leak paths, 100
leaks
bond-joint, 134–135, 138
corrosion and, 131–132, 134–135, 138
failures caused by, 131–135
gasket, 131–133, 138
materials compatibility and, 153
o-ring, 133–134, 138
in PVC piping, 136–137, 138
structural, 137–138
valve, 137
weld, 136, 138
linear thinking, 28
liquid chromatography, 105, 106(F)
litton lasers, 143
load testing, 101
log normal distribution, 161
lot changes, 85–86
lot-acceptance testing, 89, 169
probability of passing, 165–168
series tests for, 168
low-power magnification, 94
lubricant contamination, 142–143

M

magnafluxing, 93, 100, 138
magnetic particle inspection, 99, 138
mass spectrometry, 102, 104. *See also* Secondary ion mass spectrometry
mast-mounted sight (MMS), 3–5, 9
materials analysis, 94, 102–105, 106
contamination detection with, 141
of fractures, 114
of gaskets, 132
materials compatibility, 153
materials handling equipment, 142
materials strength testing, 93. *See also* Hardness testing
maximum anticipated load, 152
maximum stress, 152
McDonnell Douglas, 3–5, 54, 154–156
mean, 162, 171–172
mean squares, 173
mean time between failures (MTBF), 57–58, 60
mechanical disconnects, 118–119
mechanical failures, 152–153
mechanical inspection approaches, 93, 100–102
mechanical interference, 117–118, 141, 149–150
median, 162
meeting frequency, 72
meeting length, 26
metallic component failure
corrosion, 110–111
deformation, 115–117
fractures, 111–115, 153
mechanical disconnects, 118–119
mechanical interference, 117–118
wear, 117–118
microscopy, 93, 94–96, 129
mind-mapping, 3, 25, 27–29, 67
Monte Carlo simulation, 151

N

NASTRAN, 101
Navy, U.S., 168–169, 177
neutron beam inspection, 99–100
nonconformance documentation, 76
nonconforming components, 109
nondestructive test approaches, 93, 99–100. *See also* Neutron beam inspection; X-rays
normal distribution, 160–165
normal event, 38, 60
NORMDIST, 164–165, 170
n-rays, 93, 99, 138. *See also* Neutron beam inspection

O

operating temperature, 148–149
optical comparators, 93, 98
optical examination, 93–97, 106
optical microscopy, 93
OR gates, 38, 52, 55–56, 61–63
O-ring leaks, 133–134, 138
orthogonal arrays, 176, 179(T)
Osborn, Alex, 25
outgassing, 143
overheating, 124–125, 129
oxidation, 110

P

packaging materials, 141–142
Pareto, Vilfredo, 13
Pareto analysis, 13–15
Pareto charts, 14–15
Pareto principle, 13
pedigree analysis
  calibration data, 77
  certificates of conformance, 77–78
  drawings/specifications, 77
  inspection records, 76
  nonconformance documentation, 76
  purchase orders, 75–76
  test data, 76–77
  work instructions, 77
photography, 93, 96–97
plastic component failures, 123
plastic strain, 115
Poisson distribution, 159–160
polyvinyl chloride piping (PVC), leaks in, 136–137, 138
potential failure causes, evaluating, 72
potential solutions
  identifying, 7
  selecting best, 7–8
prepreg materials, 121–122
pressurized fluorescent dye testing. *See* Fluorescent dye pressure testing
probability
  basic failure events, 60–61
  command events, 61–63
  defined, 57
  fault-tree analysis and, 59–63
  of human error, 59–60, 66
  INHIBIT conditions, 61, 66
  MTBF and, 58–59
  normal event, 60
  top undesired event, 64–65
  undeveloped events, 61
probability distributions, 163–164
probability of passing, 165–168
problem
  defining, 2, 5, 6, 39
  identifying cause of, 2
problem-solving process, four-step, 6–8, 21, 183
process changes
  corrective, 185
  failures due to, 83–84
process guidelines, 189–190
process yield, 11
product pedigree, documents for, 75
proofreading, 22
PSpice, 151
purchase orders, 75–76

Q

Quality management, 30. *See also* Statistical process control

R

radiography, 93, 99–100, 106
Raman cell, 9, 11
redlining, 82
redundancy, 23–24, 37
redundancy defeaters, 53–55
rejection rates
  for helicopter blades, 2, 155
  hidden factory and, 12–13, 15(T)
  inspectors and, 84
  for Raman cell, 9
reliability, 165–168
"repair" disposition, 82–83
requirements flowdown, 148–149
requirements relaxation, 184
resin-based materials, failure modes of, 119
resistors
  failure of, 125
  tolerances for, 85

rework, 11, 12, 16
Rockwell hardness test, 101
rust, 110

S

safety factor, 152
sampling plan, 166
Sargent-Fletcher Company, 32–33
scales, 98
scanning electron microscopy (SEM), 95–96, 102, 103
scrap, 11, 12
secondary events, 52–53
secondary ion mass spectrometry (SIMS), 102, 104–105, 141
SEM. *See* Scanning electron microscopy
sequential AND gates, 50–51
showstopper failures, 2, 11
silicones, 143
SIMS. *See* Secondary ion mass spectrometry
single-point failures, 23–24, 37, 54
Six Sigma, 30
smart munitions, 1
smell, as burn indication, 124–125
S-N diagram. *See* Stress-number of cycles diagrams
sneak circuit condition, 21, 152
sources of change
  aging, 86–87
  corrective actions, 184–185
  design changes, 82–83, 184
  environmental changes, 84–85
  lot changes, 85–86
  process changes, 83–84, 185
  supplier, 86–87
  test/inspection changes, 84, 185
SPC. *See* Statistical process control
special actions, 186
specifications
  pedigree analysis and, 77
  system requirements in, 148
  for systems, 22
spectrographic lubrication analysis, 118
spectrometry, 102–105, 141
spectroscopy, 94, 102, 118
spring failure, 154
standard deviation, 163–164
standard tolerances, 150
state-of-part events, 52–53
state-of-system events, 52–53
statistical distributions, 157
  binomial, 158–159, 166–167
  continuous, 160–161
  discrete, 158–160
  exponential, 161
  hypergeometric, 158
  log normal, 161
  normal, 160–165
  Poisson, 159–160
  Weibull, 161
statistical experiment design
  ANOVA, 171–174, 180
  Taguchi testing, 174–181
statistical process control (SPC), 84
statistical sampling
  reliability/probability of passing and, 165–168
  use in court of, 168–169
statistical tests, 157
stress analysis, 93, 101, 111
stress fractures, 111
stress reversals, 115
stress-corrosion cracking, 111
stress-number of cycles diagrams (*S-N* diagram), 152–153
stress-strain diagram, 101
structural failures, 119
structural leaks, 137–138
sum of squares, 172
supplier, 86–87
  corrective actions by, 186
  documentation from, 76–78
  failure analysis team communication with, 8
  failure rates from, 60–61, 66
  in Ishikawa diagrams, 30
  understanding systems and, 23
surface contamination, 135
surface fractography, 94, 100, 129
surface-finish conditions, 135
surface-finish gages, 93, 98
system design inadequacy, 109
system requirements
  failure analysis and, 148
  identifying, 148–150
  operating temperature, 148–149
  relaxing, 184
systems
  definition, 19–20
  specifications for, 22
  understanding operation of, 22–23
systems failure analysis
  approach of, 5
  challenges of, 2, 5
  design/process guidelines from, 189–190
  fault-tree analysis quantification and, 65
  without hardware, 128, 129
  as part of development process, 148
  process of, 5–6
  system requirements and, 184
  troubleshooting/repair documents from, 190
systems failures
  definition, 20–21
  inadvertent system function, 21

systems failures (continued)
intermittent, 21
showstopper, 2, 11

T

Taguchi testing, 174–181
temperature
curing, 120
cycling, 102
operating, 148–149
tensile fractures, 113
tensile load, 101
test data, 76–77
test solutions, 142
testing
ANOVA, 171–174, 175(F), 180
fatigue, 93
fluorescent dye pressure, 93, 99–100, 138
goodness-of-fit, 157
hardness, 93, 100–101, 106
load, 101
lot-acceptance, 89, 165–169
materials strength, 93
nondestructive, 93, 99–100
statistical, 157
statistical sampling, 165–169
Taguchi, 174–181
vibration, 102
test/inspection changes, 84, 185
tolerance analysis, 149–150, 155
tolerances
of adhesive bonds, 119–120
determining, 150
mechanical interference/excessive clearance and, 149–150
for resistors, 85
standard, 150
torque wrenches, 93, 99
Toyoda, Sakachi, 30
Toyota Motor Company, 30
Toyota production system, 30
training, 185, 191
transfer symbols, 37, 39, 47–49
Triumph, 138–139

U

ultrasonic cleaning, 143–145
ultraviolet radiation, 102
undeveloped event, 38, 42, 43, 61, 124
uniform corrosion, 110–111
"use as is" disposition, 82–83

V

valve leaks, 137
variables data, 166
variance, 163
vibration testing, 102
vibration-induced stress failure, 101–102
vibration-induced stress reversals, 115
visual examination, 93, 124–125, 129

W

warnings, 185
warping, 153
water treatment systems, 2
WD-40, 143
wear, 117–118, 141
Weibull distributions, 161
weld leaks, 136, 138
weld porosity, 14(F), 15, 16(F), 94, 99, 136–138
wiring harnesses, 23, 37
witness marks, 117
work area contamination, 142
work instructions, 77, 83
worst-case conditions, 150, 152–153
worst-case values, 151

X

x-rays, 93, 94, 99–100, 102, 103(F), 136, 138